Microwave
Spectroscopy
of
Free Radicals

Microwave Spectroscopy of Free Radicals

ALAN CARRINGTON
Department of Chemistry
University of Southampton

1974

ACADEMIC PRESS · London · New York
A Subsidiary of Harcourt Brace Jovanovich, Publishers

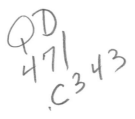

ACADEMIC PRESS INC. (LONDON) LTD.
24/28 Oval Road,
London NW1

United States Edition published by
ACADEMIC PRESS INC.
111 Fifth Avenue
New York, New York 10003

Copyright © 1974 by
ACADEMIC PRESS INC. (LONDON) LTD.

Library of Congress Catalog Card Number: 73 9451
ISBN 0 12 160750 X

PRINTED IN GREAT BRITAIN BY
William Clowes & Sons, Limited
London, Beccles and Colchester

Preface

The purpose of this book is to describe recent and current research in that part of spectroscopy which is closest to my own interests. This book is not a textbook, and it is not a comprehensive review. Rather, it is an attempt to draw together the threads of a diverse subject. The general level of the book is such that much of it can be read intelligibly by undergraduates, but since the subject itself has difficult parts, some sections of the book are tougher than others. The level at which the theory is presented is that which I consider to be the minimum worthwhile. I do not believe, for example, that microwave or radiofrequency spectra can be usefully discussed without an understanding of the concept of the effective Hamiltonian. Consequently some knowledge of quantum mechanics is essential; I have tried to indicate what the reader should know and master, however, rather than to give a course of instruction. I hope that some readers will find within this book the motivation to learn more about quantum mechanics. The experimental part of the subject is discussed in greater detail, and most of the important applications are described. I have made no attempt to be exhaustive, however, and the reference list is not to be taken as a reliable barometer of credit! I note, for example, that there is only one reference to the work of J. H. Van Vleck. Nevertheless most of the important theoretical ideas, which we now take for granted, are due to him. The experimental aspects of the subject are, in any case, advancing at such a rate that further important work will inevitably have been published before this book appears.

My initial motivation for putting pen to paper was provided by an invitation to give a course of lectures in the Chemistry Department at the University of Western Ontario. I am extremely grateful to the members of that Department for their warm hospitality during my stay. The first draft of this book was subjected to close scrutiny and constructive criticism by my own colleagues at Southampton, particularly Dr. J. M. Brown, Dr. B. J. Howard, Dr. A. C. Nelson, Dr. C. R. Parent, and Mr. R. M. Hillier. To these friends I can only express my deep thanks for their careful help. Among others who have read parts of the book and helped to improve it are Professor G. J. Hills, Mr. E. Cartmell, Dr. G. J. Lycett and Dr. R. E. Moss. Since I have freqûently been obstinate, any errors or obscurities which remain are entirely my own fault! I wish also to thank my secretary, Christine Croucher, for her great care and patience in producing the final typescript.

December, 1973   Alan Carrington

Contents

To Hilary

I.   General Introduction

I.1. Definitions

It is no easy matter to devise a title which adequately and concisely describes the contents of this monograph. Although the non-scientist may be forgiven for supposing that the language of science is always precise and unambiguous, those of us in the business know that all too often this is not true. Much of the difficulty stems from the recent development of the so-called interdisciplinary areas of science. There was a time, for example, when the term "microwave spectroscopy" had a unique meaning for most chemists and physicists. That can no longer be true. Equally, the term "free radical" was probably unambiguous to the pioneer chemists of the 1930s but now really has no precise meaning. To give just one example, the physicist studying radiation damage in potassium chloride crystals will talk about V_K centres; the chemist will talk about trapped Cl_2^- radicals. Yet other chemists will argue that "trapped" radicals, stabilized in a solid matrix, can scarcely be described as "free" radicals. All this, of course, constitutes an exercise in semantics, but the difficulty is none the less real and the first task of an author must be to describe the scope of his interest, and to attempt to define the sense in which he uses particular terms.

We therefore deal first with the chemical aspects and discuss the sense in which the term "free radical" is used. In this book we shall be dealing with three types of chemical species, namely, molecules with open-shell ground electronic states, molecules in either open or closed-shell excited electronic states, and a few molecules which, although they have closed-shell ground states, nevertheless have only a transitory existence under normal conditions. The first category includes most of the species which the majority of chemists would describe as free radicals, species such as SO, HCO, ClO, etc. Many of them possess electrons with unpaired spin and therefore exhibit large magnetic

1

moments. But the category, as stated, also includes stable bottled gases like O_2 and NO. Oxygen, of course, has an open-shell electronic structure with two unpaired electrons and although it would not normally be classed by chemists as a free radical, the species SO which has a similar electronic structure undoubtedly would. Here the distinction is essentially one based on considerations of chemical reactivity; we shall not, however, concern ourselves with such a distinction because a discussion of the spectroscopic properties of SO which did not also refer to O_2 would be absurd. Similarly, the electronic structures and spectroscopic characteristics of NO and ClO have much in common, as also have NO_2 and HCO.

The second category, molecules in open- or closed-shell excited electronic states, would, it must be admitted, rarely be classified as free radicals. Nevertheless they are frequently highly reactive and short-lived, and in their electronic structures they share many of the characteristics of the members of our first category. Some of the most beautiful and novel recent experiments in microwave spectroscopy have been concerned with excited electronic states, and that is a good enough reason for their inclusion here.

Finally, our discussion would be incomplete if it did not include molecules like CF_2 and SiF_2, which have closed-shell ground states. These would surely be classified as free radicals by most chemists, because they are relatively short-lived under normal circumstances. Nevertheless many features of their structure and microwave spectra are similar to those of H_2O, certainly not a free radical by anyone's definition!

Similar difficulties exist with the term "microwave spectroscopy". The experiments described in this monograph involve the use of electromagnetic radiation from frequencies as low as a few megahertz (i.e. wavelengths about 300 metres) to frequencies up to several hundred gigahertz (wavelengths about 1 millimetre). Thus the range extends from the radiofrequency region, through the microwave region, to the far infrared, although most of the work described falls into the microwave region. We also use the term "microwave spectroscopy" to cover any experiment in which the absorption or emission of microwave radiation by a molecular species can be detected directly or indirectly. The molecule may, or may not, be additionally perturbed by other electric or magnetic fields. This apparently pedantic statement is, unfortunately, necessary because there are those to whom the term "microwave spectroscopy" means only the *direct* detection of the *absorption* of microwave radiation, in the absence of other electromagnetic fields, apart from modulating fields.

Finally in defining the title of this book it is necessary to add that only experiments on molecules in the gaseous phase at relatively low pressures are discussed. Of course, the term "microwave spectroscopy" can be used legitimately to cover, for example, electron spin resonance studies of liquid phase free radicals. However most spectroscopists would agree that, particularly for microwave studies, there are fairly sharp distinctions to be made when comparing the gaseous with the condensed phases. In the gas phase at low pressures (i.e. a few torr or less) the rotational motion of the molecules is quantized whereas in the liquid phase it is randomized by the frequent molecular collisions, and in the solid phase is usually suppressed totally. In most gas phase investigations the interactions between molecules represent relatively small perturbations which usually affect only the widths of the spectral lines; in molecular beam studies such perturbations are absent. In contrast the very strong intermolecular interactions in condensed phases can be dominating influences on the spectral characteristics and, indeed, spectroscopy is a very powerful tool for investigating intermolecular interactions. There are, of course, many molecules which are only weakly perturbed by their environment in condensed phases, and in such cases valuable comparisons of results obtained in different phases can be made, as we shall show. But there are others where such comparisons must be approached with considerable caution. Nitric oxide in the gas phase, for example, is totally different from nitric oxide trapped in a solid matrix. We will have more to say on this subject elsewhere.

With these preliminary remarks in mind we now turn to a brief outline of the main experimental techniques to be described in greater detail in Chapter II.

I.2. Microwave Spectroscopy

I.2.1. Introduction

The object of a spectroscopic experiment on a molecule is to measure the relative positions of two or more energy levels by allowing the molecule to interact with electromagnetic radiation and determining the frequency of the radiation absorbed or emitted during the interaction process. We can divide the techniques used into direct and indirect methods. In the former the absorption or emission of radiation is detected directly by monitoring the total radiation power level. In the latter we detect the change in the molecule induced by the radiation by monitoring some *other* property of the molecule, for example, its radiative emission at some quite different frequency, or a change in its translational motion. The direct methods are conceptually the simplest, but not necessarily the easiest to put into practice or the most sensitive. We deal with them first.

I.2.2. Direct methods

We begin by considering two adjacent energy levels in a molecule, the separation corresponding to a frequency in the microwave region; they might, for example, be two rotational levels in a diatomic molecule, K-doublet components in an asymmetric rotor, spin doublets in a doublet state radical, etc. For an assembly of molecules in the gas phase a certain number N_1 will be in the lower state, and usually a different number N_2 in the upper state; we label the states simply 1 and 2. Since molecules actually possess very many different energy levels, most of the molecules in a macroscopic sample will be in some other unspecified state, but we are not here concerned with them. So far as states 1 and 2 are concerned we can visualize four different limiting cases, illustrated in Fig. I.1. In the first, case a, the relative populations are determined by Boltzmann's distribution law. If we are not perturbing the molecules in some way, and the pressure of the gas is high enough for frequent molecular collisions to occur, the ratio of the populations is given by the simple equation,

$$\frac{N_2}{N_1} = \exp - \left(\frac{E_2 - E_1}{kT} \right) \qquad (I.1)$$

where E_1 and E_2 are the energies of the states, T is the absolute temperature of the bulk gas, and k is Boltzmann's constant.

The other three cases present unusual situations which can be achieved in the laboratory by perturbing the system. Case b, where the populations N_1 and N_2 are equal is known as a *saturation* condition, case c with $N_2 \gg N_1$ represents a *population inversion*, whilst case d with $N_1 \gg N_2$ will be called a *population cooling* since it corresponds to the situation described by equation

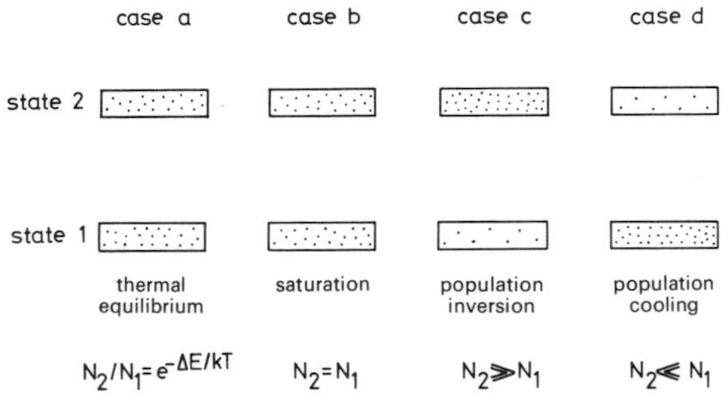

Fig. I.1. Limiting populations for two energy levels in a macroscopic sample.

(I.1) with T at or near absolute zero, even though we may, in other respects, suppose the system to be at a normal temperature (room temperature in Britain, somewhat below room temperature in the U.S.). We will see that all four situations (and, of course, all the intermediate situations) can be realized and lead to different spectral characteristics.

Next we must consider the ways in which transitions between the two states can occur, and here we can identify three different possibilities, as follows:

(i) *Transitions induced by collisions between molecules or between molecules and the walls of the containing vessel.* In an idealized example, molecule A in state 1 collides with molecule B in state 2, with the result that A emerges in state 2 leaving B in state 1. In practice, of course, molecules can arrive at states 1 or 2 from other unspecified states. The important point is that these collision-induced transitions, which constitute the so-called collisional relaxation process, establish and maintain the Boltzmann distribution given by equation (I.1). If we achieve artificially population cases b, c or d, we can only maintain them either by competing effectively with the collisional relaxation, or by removing the possibility of collisions, as in a molecular beam experiment.

(ii) *Spontaneous transitions from the upper to the lower state, with the emission of radiation.* The probability, A, of such a transition was given by Einstein as

$$A = \frac{16\pi^3 v_{12}^3}{3\varepsilon_0 hc^3} |\mu_{12}|^2 \quad \text{(in SI units)} \tag{I.2}$$

where v_{12} is the frequency of the emitted radiation, ε_0 is the permittivity of free space, h is Planck's constant, c is the velocity of light, and μ_{12}, called a transition matrix element, is a quantity which is characteristic of the molecule and the type of transition, the nature of which we shall discuss at length later. For levels whose separation corresponds to a radiofrequency or microwave frequency the spontaneous emission process is very slow. When the frequency v_{12} corresponds to the infrared, visible or ultraviolet regions of the spectrum, however, spontaneous transitions are highly probable and the emitted radiation is well known as fluorescence, phosphorescence or chemi-luminescence.

(iii) *Transitions induced by interaction of the molecule with electromagnetic radiation.* A molecule can interact with electromagnetic radiation by virtue of either its electric or magnetic moments. Electric dipole transitions are, as the name suggests, induced by coupling of the oscillating electric field component of the radiation with the electric dipole moment, permanent or induced, of the molecule. Although higher electric moments can also couple with radiation, transitions induced by such coupling are relatively rare and will not be mentioned again. The electric dipole transitions we shall be interested in are invariably induced by coupling of the permanent electric dipole moment of the molecule with the radiation field. Magnetic dipole transitions, on the other hand, are induced by interaction of the oscillating magnetic component of the radiation with a magnetic dipole moment of the molecule. The latter may arise from unpaired electron spin, nuclear spin, the molecular dipole moment caused by rotation of the molecule, etc. The Einstein B coefficient governs the probability of a radiation-induced transition and is given by

$$B = \frac{2\pi^2}{3\varepsilon_0 h^2} |\mu_{12}|^2. \tag{I.3}$$

The matrix element μ_{12}, in fact, describes the nature and magnitude of the coupling of the molecule with the radiation. A particularly important point to note is that the probabilities for induced emission and absorption are the same, i.e.,

$$B_{21} = B_{12} = B. \tag{I.4}$$

In a spectroscopic experiment we measure the intensity of the radiation emitted or absorbed by an assembly of molecules. For a spontaneous emission process the power intensity of emitted radiation, I_{21}, is given by

$$I_{21} = N_2 h v_{12} A \quad (\text{J s}^{-1}) \tag{I.5}$$

where $h v_{12}$ is the energy of each quantum of radiation. Substitution for the A coefficient using equation (I.2) gives the result

$$I_{21} = \left(\frac{16\pi^3}{3\varepsilon_0 c^3}\right) N_2 v_{12}^4 |\mu_{12}|^2. \tag{I.6}$$

Note the dependence on the fourth power of the frequency, which explains why, despite its relative unimportance at microwave frequencies, spontaneous emission becomes an important process at infrared, visible and ultraviolet frequencies. At 1 cm wavelengths with an electric dipole transition, μ_{12} is typically of the order 3×10^{-30} C m and A is therefore of the order 10^{-7} per second.

In the case of radiation-induced transitions we must take account of the density of the radiation at frequency v_{12}, which we denote as $\rho(v_{12})$. The probability of induced absorption, P_{12}, is given by

$$P_{12} = \rho(v_{12})B \tag{I.7}$$

whilst the total probability of induced and spontaneous emission, P_{21}, is given by

$$P_{21} = \rho(v_{12})B + A. \tag{I.8}$$

In order to complete the description of the system, the effects of collisional relaxation must be included. This may be done phenomenologically by treating such relaxation as arising from radiation components of $\rho(v_{12})$ generated by the collisions.

From (I.7) and (I.8) it is a simple matter to write down expressions for the number of molecules making upwards and downwards transitions in a small time interval Δt,

$$N_{1\to 2} = N_1 B\rho(v_{12})\,\Delta t, \qquad N_{2\to 1} = N_2[B\rho(v_{12}) + A]\,\Delta t \tag{I.9}$$

and since at thermal equilibrium these numbers must be equal, we obtain the ratio

$$\frac{N_2}{N_1} = \frac{B\rho(v_{12})}{B\rho(v_{12}) + A}. \tag{I.10}$$

Comparing this with equation (I.1) we can derive the results for thermal equilibrium,

$$\rho(v_{12}) = \frac{A}{B}\frac{1}{e^{hv_{12}/kT} - 1} \tag{I.11}$$

which when compared with Planck's law for the energy density of a black body emitting radiation,

$$\rho(v_{12}) = \frac{8\pi h v_{12}^3}{c^3}\frac{1}{e^{hv_{12}/kT} - 1} \tag{I.12}$$

leads directly to a relationship between the Einstein A and B coefficients,

$$A = \frac{8\pi h v_{12}^3}{c^3} B. \tag{I.13}$$

Now let us return to the experiment and suppose that $\rho(v_{12})$ represents the density of monochromatic radiation from a microwave source such as a klystron or backward-wave oscillator. Then the rate of absorption of energy is

$$I_{12} = N_1 B\rho(v_{12})hv_{12} \tag{I.14}$$

whilst the power returned by stimulated emission is

$$I_{21} = N_2 B \rho(v_{12}) h v_{12}. \tag{I.15}$$

For the reasons given earlier we can neglect the very small amount of spontaneously emitted power. Combining (I.14) and (I.15) we see that the net change is

$$\Delta I = I_{12} - I_{21} = (N_1 - N_2) B \rho(v_{12}) h v_{12}. \tag{I.16}$$

We are now in a position to examine the four population cases illustrated in Fig. I.1, using equation (I.16) to investigate the possible spectroscopic experiments which may be performed.

(a) In case a where N_1 is greater than N_2 according to the Boltzmann distribution law, ΔI is positive and we observe a net absorption of power. Substituting the Boltzmann distribution into (I.16) we obtain for the absorption intensity

$$I_{abs} = \Delta I = N_1 [1 - e^{-h v_{12}/kT}] B \rho(v_{12}) h v_{12} \tag{I.17}$$

and since at microwave frequencies and normal temperatures, $h v_{12} \ll kT$, we can expand the exponential as a convergent series to yield

$$I_{abs} = \frac{N_1 (h v_{12})^2}{kT} B \rho(v_{12}) \left[1 - \frac{1}{2} \frac{h v_{12}}{kT} + \cdots \right] \tag{I.18}$$

$$\simeq \frac{N_1 (h v_{12})^2}{kT} B \rho(v_{12}). \tag{I.19}$$

We shall develop this expression further when we go into the experimental details in the next chapter. An important point to note is that provided a Boltzmann distribution is maintained the absorption intensity increases as the square of the transition frequency.

(b) In the population case b where $N_1 = N_2$, equation (I.16) tells us that $\Delta I = 0$ so that there is no change in the radiation power level. This condition, known as saturation, can be achieved by increasing $\rho(v_{12})$ to the point where the radiation-induced transition rate dominates the collisional relaxation rate. In a simple absorption experiment, therefore, it is important to keep the power level low enough to avoid the onset of saturation.

(c) For population case c where we have an inversion ($N_2 \gg N_1$), equation (I.16) shows that ΔI is negative and stimulated emission of radiation will be observed. The snag is that in order to obtain a population inversion and maintain it, the effects of molecular collisions must be overcome and even then the radiation-induced transitions can lead to equalization of the populations. One way of solving the problem is by using a molecular beam in which the radiation is continuously presented with a fresh supply of molecules in state 2. The population inversion is usually achieved by first passing the beam through an electrostatic state selector which focusses upper state molecules but rejects the lower state. The result of the experiment is essentially an amplification of the incident microwave radiation and the first successful device was termed a "maser' by its inventors.

(d) Case d is similar to case c and would lead to the observation of enhanced stimulated absorption. Molecular beam techniques are again useful but now require a state selector which focusses molecules in the lower state 1. The experiment has been performed successfully, but is not yet widely used.

In all the above experiments, which are summarized in Fig. I.2, the spectrometer is usually designed so that the source frequency is swept through the value v_{12} required for the transition. The spectrum is thus presented as an absorption or emission of microwave radiation, plotted as a function of frequency. The details of how this is done will be described in the next chapter; the simple absorption experiment (a) will be described as "microwave rotational spectroscopy" since the energy levels involved are usually adjacent rotational levels. Experiment (c) will, of course, be described as molecular beam maser spectroscopy.

Before leaving the direct methods, however, we must describe an important extension of the techniques discussed above. The energy states 1 and 2 may be stationary states of the unperturbed molecule, but equally they may be states whose separation is determined in whole or in part by a perturbing electric or magnetic field. For example, they could arise from interaction of an unpaired electron spin with an applied magnetic field. If the electron spin is decoupled

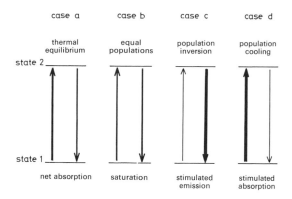

Fig. I.2. Consequences of stimulated transitions for the four population cases described in Fig. I.1.

from all other angular momenta, the separation of the states is essentially $2\mu_B\mathbf{B}$, where μ_B is the electron Bohr magneton and $\mathbf{B}$ is the strength of the applied magnetic field; in zero field ($\mathbf{B} = 0$), the two states are actually degenerate. Our analysis still applies and the simple absorption experiment corresponding to case a is known as electron spin resonance. Alternatively the two states could be rotational levels of a diatomic molecule with $S = \frac{1}{2}$ in which the spin angular momentum is strongly coupled to other angular momenta, and which is again perturbed by an external field. Then again, they might be spatial components of the total angular momentum of a closed-shell molecule that have been separated by an applied electric field. Many such situations are possible and will be discussed extensively elsewhere. The main point to be stressed now, however, is that since the frequency separation is dependent upon the strength of a static magnetic or electric field, a spectroscopic experiment can be carried out using a *fixed* frequency but varying the applied field. Transitions then occur when the separation between the states is adjusted to the frequency of the radiation. Many absorption experiments on gaseous free radicals have been performed and they are usually described as "gas phase electron resonance" experiments because the magnetic tuning of the levels is due to the presence of electronic angular momentum, spin or orbital, within the molecule. Equally well, molecular beam maser experiments can also be carried out using a fixed microwave frequency but with a variable applied electric or magnetic field.

I.2.3. Indirect methods

The indirect methods which have been used to study free radicals and excited states fall into two distinct classes, namely, molecular beam deflection experiments and double resonance methods. These are different in concept and execution and we therefore discuss them separately, dealing first with the molecular beam methods.

A molecular beam is formed by allowing molecules to effuse through a suitable orifice into a chamber which is maintained at a very low pressure (10^{-6} torr or less). Beam sources are designed to produce as high a degree of forward directionality as possible, and those molecules which are not travelling in the desired direction through the apparatus are pumped out to avoid scattering problems. We will describe the techniques used in producing molecular beams in more detail in the next chapter, but for the moment assume that a reasonably intense narrow beam can be formed and its flux density monitored continuously.

Molecular beams can be deflected by passing them through suitable magnetic or electric fields. Let us return to the simple two level system described in the previous section and suppose that the two states are of opposite parity, so that in an electric field the energy of the upper state increases whilst that of the lower state decreases. If the beam is passed through a quadrupole electrostatic field, molecules in the upper state are focussed but those in the lower state are rejected. The state-selected beam can then be passed through a dipole deflecting field which directs the beam to a detector giving an output proportional to the beam flux density. If between the quadrupole and dipole fields, transitions between the two states are now induced by a radiofrequency or microwave field, some of the upper state molecules are converted to the lower state, do not satisfy the dipole focussing condition, and therefore do not reach the detector. Consequently absorption of microwave power by molecules in the beam causes a reduction in beam intensity and hence a spectrum can be obtained. Figure I.3 illustrates a typical experiment. The first deflecting or focussing field is called the A field, whilst the last

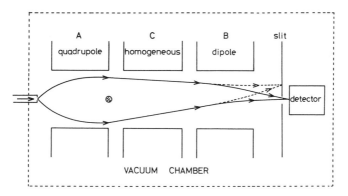

Fig. I.3. Typical molecular trajectories in a molecular beam electric resonance experiment.

refocussing field is called the B field. The transition region, in which the molecules are usually subjected to both an oscillating (radiofrequency or microwave) and a static electric field, is called the C field region. Typical trajectories of two molecules are illustrated; the solid lines indicate the trajectories of upper state molecules which are allowed to reach the detector, whilst the dotted lines show the trajectories *after* upper state molecules have been converted to the lower state and are then unable to reach the detector. Since the spectroscopic transition causes a reduction in beam intensity at the detector, it is known as a "flop-out" experiment. One could alternatively arrange matters so that only lower state molecules reach the detector, in which case the spectroscopic transition produces an increase in beam intensity at the detector (this is called a "flop-in" experiment). The detector could be a mass spectrometer, but is also often an Auger detector, which is a particularly sensitive detector for molecules in excited *electronic* states.

In practice, the energy levels of interest usually have spatial degeneracy which is removed by an electric field, leading to splitting into separated components. Nevertheless the principles of the experiment remain the same, in that states whose energy increases are selected, whilst those whose energy decreases are rejected. There will, however, be more than one spectroscopic transition and hence more than one absorption line. We will go into the details of several such experiments in the next chapter. Magnetic focussing and deflecting fields can also be used, particularly for species in open-shell electronic states. In such cases the C field will also be a magnetic field and the transitions studied are similar to those in the electron resonance experiment.

It is clear that molecular beam deflection techniques are of considerable generality and high sensitivity. Up to the present time they have been mainly used for excited electronic states because of the high efficiency of surface ionisation detectors. With advances in mass spectrometric detection, however, it may soon be feasible to apply the method more readily to open-shell ground state molecules.

We now turn to the second class of indirect methods, the so-called "double resonance" experiments which have been elegantly devised to study molecules in excited electronic states. Consider the simple idealized situation illustrated in Fig. I.4. E_2 and E_3 might be adjacent *rotational* levels (with separation ~ 1 cm^{-1}) in an excited *electronic* state ($\sim 10^4$ cm^{-1} above the ground state); E_1 is a particular rotational level of, say, positive parity in the ground electronic state. Suppose that excitation of the molecule from E_1 to E_3

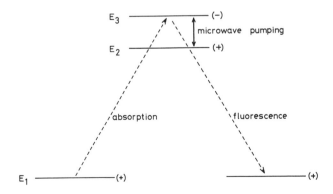

Fig. I.4. Principles of a microwave/optical double resonance experiment.

can be accomplished by absorption of light in the visible or ultraviolet regions of the spectrum. Suppose also that the exciting light is coincident with the energy separation between E_1 and E_3, but does not excite any other transitions. The molecule can then return from E_3 to the ground state, either by non-radiative processes, or by the emission of light (fluorescence). Although fluorescence transitions to different rotational and different vibrational levels of the electronic ground state occur, we choose to monitor only the E_3 to E_1 transition. If now the transition between E_2 and E_3 is excited with intense microwave radiation of the correct frequency, molecules in the E_3 state, of negative parity, will be converted into E_2 state molecules of positive parity. Since the fluorescence transition from E_2 back to E_1 is electric dipole forbidden (because both states have the same parity), a decrease in fluorescence intensity will be observed. Hence the microwave transition $E_2 \leftrightarrow E_3$ in the excited electronic state can be detected by continuous monitoring of the fluorescence, which decreases in intensity as the microwave frequency is swept through the necessary value. The method, which is called microwave/optical double resonance, is extremely sensitive since the fluorescent radiation can be detected with high efficiency.

In practice most experiments are considerably more complicated, and many variations are possible. Excitation from the ground to the excited electronic state may be accomplished with electromagnetic radiation, by electron bombardment, or even by chemical reaction. It is not usually possible to isolate a single fluorescence line, and when several lines are detected simultaneously, the measurements and their interpretation are not so clear cut. In other experiments, the levels may be perturbed by electric or magnetic fields, in which case the effect of the microwave "pumping" may be a change in the polarization of fluorescence. We will meet a number of different examples in the next chapter.

Finally a remark about the term "double resonance" which is usually used to describe such experiments. The word "resonance" was originally introduced in connection with the classical precession model of an electron or nuclear spin in a magnetic field. According to this model, the spin vector precesses about the direction of an applied static field **B** with a characteristic precession frequency, and an oscillating magnetic field applied perpendicular to **B** may also be decomposed into components precessing about **B**. When the precession frequencies for the spin and magnetic field vectors are the same, they come into "resonance" and energy is absorbed. In terms of quantized energy levels, a spectroscopic transition is induced. The current use of the term "resonance" to describe almost any radiofrequency or microwave experiment, even when static fields are absent, is in some ways unfortunate but it can be justified and is now so widespread that we must learn to live with it.

I.3. Other Techniques

The remaining sections of this chapter are devoted mainly to the questions of why the microwave spectra of free radicals are of interest, and what information they provide which is not otherwise available. Consequently it is worth considering the other spectroscopic techniques used in studying free radicals, and of these, the most important are electronic (visible and ultraviolet) spectroscopy of gaseous species, electron spin resonance studies in condensed phases, and infrared work, mainly on radicals trapped in solid matrices.

One of the most significant advances in the spectroscopy of free radicals was the invention of flash photolysis, in which free radicals are produced by irradiation of a gaseous sample with an intense flash of light, and their absorption or emission spectra photographed a very short time later. Photographic techniques in the visible and ultraviolet are ideally suited to situations in which rapid recording is essential, and the flash photolysis method makes it possible to observe radicals with lifetimes in the microsecond range. The spectra provide information about ground and excited electronic states and the resolution obtainable leads, through the measurement of vibrational and rotational fine structure, to fairly precise information about molecular geometry. The limiting factor, however, is the line width which, as we shall see, is usually determined by the Doppler effect. In general the line widths obtainable in the optical region are approximately 10^4 times greater than can be achieved in the microwave region using conventional methods, and 10^6–10^7 times larger than is achievable with molecular beam methods. Hence many of the small intramolecular interactions, and the perturbations produced by external fields, which are so informative about the electronic structure of the species are inaccessible to the electronic spectroscopist.

Infrared and Raman methods of studying free radicals have, on the whole, been less successful. In the gas phase only long lived species have been detected until very recently and most of the work described has been concerned with radicals trapped in solid matrices, usually at low temperatures. A knowledge of the vibration frequencies is, of course, of considerable value in helping to establish molecular geometry, but the perturbing effects of the environment must always be taken into account. The resolution obtainable is no better than that realized in the visible and ultraviolet regions of the spectrum, and since vibrational frequencies are usually obtainable from electronic spectra, solid state vibrational spectroscopy of free radicals does not, on the whole, compete too successfully with gas phase electronic or microwave spectroscopy. Gas phase vibrational spectroscopy of free radicals would be exciting, but the technical difficulties have yet to be overcome although some very recent work on the methyl radical is a major step forward.

Electron spin resonance studies of free radicals in solids and liquids have been of immense value in contributing to our knowledge of the structure of free radicals. The resolution obtainable is similar to that achieved in conventional gas phase microwave spectroscopy, and many of the intramolecular interactions which we shall discuss elsewhere in this book can and have been investigated by electron spin resonance. The major difference, of course, is that the quantized rotational motion of molecules is not conserved in condensed phases; hence *direct* determinations of molecular geometry are not possible, although solid state studies do usually establish the site symmetry, from which molecular symmetry can be inferred. Furthermore, the fact that many of the intramolecular interactions and perturbations produced by external electric and magnetic fields are anisotropic means that solid state studies, particularly single crystal measurements, are of particular significance. Measurements of the anisotropy lead directly to knowledge of the electronic structure and symmetry and, by inference, to molecular geometry. We shall find it illuminating to compare gas phase results with solid state studies of the same molecular species, but we shall also see that the perturbing effects of the environment in the solid state lead to complications which must be fully appreciated.

In the liquid phase the rapid random Brownian motion of the molecules leads to complete averaging of all anisotropic interactions, and only the isotropic effects are available for measurement. Although in one sense this may be regarded as an unfortunate loss of information, it does also mean that much larger and more complicated molecules are accessible for study. One of the main applications of liquid phase electron spin resonance has been to the study of large organic free radicals which, despite the removal of anisotropic interactions, exhibit complex spectra due to interactions between electron and nuclear spins. If the gas phase spectra of such species could be obtained they might well be too complex to be resolvable and would certainly present formidable problems of analysis. Even a spectroscopist is willing to admit that a spectrum may occasionally contain too many lines!

It would be a mistake to suppose that these different forms of spectroscopy are in competition with each other, even if their practitioners sometimes appear to be so. The techniques are complementary, and the information from one is often crucial to the successful application of another. For example, one could scarcely attempt a microwave/optical double resonance study without the knowledge of both the ground and excited states provided by electronic spectroscopy. So although this book is about microwave spectroscopy, we shall often find it valuable to refer to other spectroscopic studies.

I.4. Line Widths and Resolution

All spectroscopists strive constantly for higher sensitivity and higher resolution. It is time we looked for the reasons why microwave techniques offer high resolution, and we therefore now consider the mechanisms of line broadening and assess their relative effects in the different microwave experiments to be described.

I.4.1. Natural line width

We have already noted the importance of spontaneous emission processes as a source of radiation, particularly at visible and ultraviolet wavelengths. Spontaneous emission, however, also makes a contribution to the line width, and in some instances even dominates the experimental observations. The origin of this contribution is that spontaneous emission terminates the life of the upper state in a given molecule, and the Heisenberg uncertainty principle states that the lifetime of the state (Δt) and uncertainty in its energy (ΔE) are related by the expression

$$\Delta t . \Delta E \approx \hbar \tag{I.20}$$

where $\hbar$ is Planck's constant divided by 2π. The corresponding spread in frequency Δv is

$$\Delta v = \frac{\Delta E}{h} \approx \frac{1}{2\pi\,\Delta t} \tag{I.21}$$

so that we must determine Δt. This is straightforward since Δt is simply equal to the inverse of the Einstein A coefficient,

$$\Delta t = \frac{1}{A} = \frac{3\varepsilon_0 hc^3}{16\pi^3 v^3 |\mu|^2}. \tag{I.22}$$

Consequently the frequency spread, and hence line width, of a spectroscopic transition to or from the state in question, is

$$\Delta v = \frac{8\pi^2 v^3}{3\varepsilon_0 hc^3} |\mu|^2 \approx 1{\cdot}66 \times 10^{20}\, v^3 |\mu|^2. \tag{I.23}$$

If we insert a typical value of 3×10^{-30} C m for the matrix element μ, and put $v = 10^{10}$ Hz, equation (I.23) predicts a natural line width of $\sim 10^{-9}$ Hz. We shall see that, in the microwave region of the spectrum, this contribution to the line width is negligible in comparison with other contributions. In the visible and ultraviolet regions, however, the radiative line width is often by no means negligible; at a wavelength of 4000 Å, Δv is about 1 MHz. If, however, the excited electronic state is metastable, μ may be very small and Δv is then also very small.

I.4.2. Pressure broadening

We showed in section I.2 that conventional microwave absorption experiments can only be performed successfully if the populations of the levels involved are maintained at or near their thermal Boltzmann equilibrium values. We indicated that molecular collisions are responsible for the preservation of thermal equilibrium by inducing non-radiative transitions between the levels. It follows from section I.4.1, however, that if the state lifetimes are limited by molecular collisions, so also is the line width. At pressures in the region of 1 torr it is only necessary to consider two-body collisions, and the line width for most gases due to collisions at this pressure is in the range 1–10 MHz.

The theory of pressure broadening has been very fully developed because it provides information about the interactions between molecules. Not all collisions necessarily result in non-radiative transitions and, conversely, it is not always essential for two molecules to actually collide head-on in order for exchange of energy to occur, long-range interactions frequently being important. The aim of theoretical and experimental studies of pressure broadening is to identify the main interactions which induce radiationless transitions, and hence to determine molecular constants (like molecular quadrupole moments) which are not readily obtainable by direct measurements. We shall not pursue the subject further since in the free radical field, molecular collisions can have the more disastrous result of removing the species entirely through chemical reaction, or in the case of excited electronic states, by inducing radiationless electronic transitions to the ground state. For these reasons free radical studies are usually conducted at the lowest practical pressures and, up to now, pressure broadening studies have not aroused much interest.

I.4.3. Doppler broadening

Most readers will be familiar with the Doppler effect in acoustics; if a moving vehicle is emitting a note of constant frequency, the note heard by a stationary observer depends upon the velocity of the vehicle and, in particular, upon whether the vehicle is moving towards or away from the observer. Exactly the same effect is important in spectroscopy: the molecule is the moving vehicle, emitting (or absorbing) radiation, and the receiver system of the spectrometer is the observer.

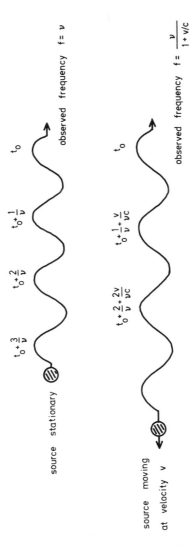

Fig. I.5. The Doppler effect and resulting frequency shift.

We can express the Doppler effect quantitatively with the help of Fig. I.5. Consider first a stationary source emitting radiation of constant frequency v. Suppose that the radiation is sinusoidal and the observer marks the time at which each wave node passes. Clearly if the first node occurs at time t_0, successive nodes will be counted at times $t_0 + 1/v$, $t_0 + 2/v$, etc. In other words, the observed frequency f is the same as the emitted frequency v, and the observed wavelength λ is equal to c/v.

Now suppose that the source is moving away from the observer at velocity v. The observer notes the first node at time t_0, but because the source is moving away, the second node takes a little longer to pass the observer than before and is noted at time $t_0 + 1/v + v/vc$. Successive nodes occur at times $t_0 + 2/v + 2v/vc$, $t_0 + 3/v + 3v/vc$, etc. Hence in this case the observed frequency f is given by

$$f = \frac{1}{\{1/v + v/vc\}} = \frac{v}{\{1 + v/c\}} \tag{I.24}$$

and the observed wavelength is

$$\lambda = \frac{c}{f} = \frac{v + c}{v}. \tag{I.25}$$

In the case of a source moving towards the observer (v negative), the observed wavelength decreases (v increases), whilst for a source moving away from the observer (v positive), the observed wavelength increases (frequency v decreases). The Doppler shift z is given by the equation

$$z = \frac{\lambda - \lambda_0}{\lambda_0} = \frac{v}{c} \tag{I.26}$$

where λ_0 is the emitted wavelength, provided $v \ll c$, which is usually the case for molecules.

In practice, of course, spectroscopic investigations usually involve the study of an assembly of molecules moving in different directions at different velocities, and suffering frequent collisions which change both the velocity and direction. Each molecule exhibits a different Doppler shift and hence an absorption line (or emission line) from an assembly of molecules is not shifted but broadened. It may be shown that for a gas at thermal equilibrium, the line width Δv at half-height is given by

$$\Delta v = \frac{2v}{c}\left(\frac{2NkT \ln 2}{M}\right)^{\frac{1}{2}} = 7{\cdot}15 \times 10^{-7}(T/M)^{\frac{1}{2}} v \tag{I.27}$$

where N is Avogadro's number and M is the mass of the molecule. As an example, put $T = 300$ K and $M = 30$; we then calculate that for a microwave frequency v of 10^{10} Hz, Δv is about 23 kHz. In the ultraviolet region of the spectrum (say $\lambda = 3000$ Å, for which $v = 10^{15}$ Hz), however, the line width given by (I.27) is 2300 MHz. Molecular beams, in which all the molecules are moving in the same direction, represent a special case which we will discuss later. Interstellar astronomy is also a special case in that an interstellar gas cloud emitting radiation is invariably moving relative to the earth, and in this case a Doppler shift is observed.

I.4.4. Instrumental effects

Finally we consider briefly a number of instrumental effects which can contribute to or even determine the line width in a microwave experiment. In principle they are avoidable, but in practice the technical difficulties are often considerable.

First among these is saturation, that is, equalization of the populations of the levels involved through the use of too much microwave power, as discussed in section I.2. The main result of saturation is a reduction of the peak height in an absorption experiment, but since the reduction is greatest at the centre of the absorption line and less important in the wings, the line becomes both weaker and broader.

Collisions of the molecules with the walls of the containing cell can also determine or contribute to the line width when the pressure is low enough for the mean free path of the molecules to be greater than the dimensions of the cell. This effect is normally only important when very high frequency measurements are being made at very low gas pressures.

It is clearly necessary that the microwave source have a frequency stability considerably better than the spectral line width, and in experiments involving the application of static electric or magnetic fields, the field homogeneity must be high enough to avoid distortion of the line shape. Lastly it is important that modulation and receiver systems be designed so as to detect and display the true line shape without distortion.

I.4.5. Limiting line widths in different experiments

In any particular microwave experiment one or more of the effects listed may contribute to the total line width and the task of the spectroscopist is usually to identify the major contribution and attempt to reduce it. In conventional microwave absorption experiments and in gas phase electron resonance, pressure broadening dominates at pressures of the order of 1 torr, but if the pressure is reduced, as it usually can be, pressure broadening can be removed entirely. The line width is then usually dominated by the Doppler effect, which is not so readily reduced. In molecular beam experiments the absence of molecular collisions means that pressure broadening is absent, and since the direction of propagation of the microwave radiation and direction of flow of the molecular beam are usually arranged to be perpendicular, Doppler broadening is also removed. In many experiments the width-determining consideration is the time (Δt) the molecules spend in the microwave radiation field (by equation (I.20)). The value of Δt, in turn, depends on the average molecular velocity in the beam, and the length of the radiation field. The latter can be increased to the extent that line widths down to even a few hertz can be achieved in some special cases. Certainly the resolution of beam techniques cannot be approached with the other microwave techniques described in this book, although it should be mentioned that beam experiments with applied static electric or magnetic fields are, in practice, often restricted by field inhomogeneities. The microwave/optical double resonance methods are also capable of yielding very narrow lines but in many instances radiative decay of the excited electronic state to the ground state limits the excited state lifetime, and hence dominates the microwave line widths.

Finally in comparing microwave and optical methods, it is worth noting that line widths in the visible and ultraviolet regions are usually limited by the Doppler broadening, which at such very high frequencies is rather large, as we have seen. Special techniques for overcoming this Doppler broadening are being developed, but it is not yet clear that they will be of general applicability. It seems likely that, for the foreseeable future, optical spectroscopy will not compete effectively with microwave techniques so far as spectral line widths are concerned. But to repeat an earlier remark, the techniques are complementary and not in competition. Each provides information which is unique and important.

I.5. Application of Free Radical Microwave Studies

In concluding this introductory chapter it is pertinent to ask (and answer) what is the justification for the effort involved in detecting and measuring the microwave and radiofrequency spectra of free radicals. After all, free radicals are in a sense, molecular freaks, since most molecules have closed shell electronic ground states and are long-lived under normal conditions. Does it not follow that the spectroscopy of free radicals is a backwater, fascinating mainly because it is difficult?

There are, of course, a number of different answers to these questions. To start with a generalization, it might well be maintained that in many fields of science we learn more by studying the abnormal rather than the normal. In particular, much of our knowledge of the electronic structure of molecules has been derived from spectroscopic studies of free radicals. The main reason for this, and the reason why free radicals are of particular interest to spectroscopists, lies primarily in their open-shell electronic structures. The presence of electrons with unpaired spins means the existence of magnetic interactions, both within the molecule and with external fields, which are not present with closed shell molecules. These interactions, which we shall discuss in detail in Chapter III, lead to complex spectra whose successful analysis often provides intimate details of the molecular structure. The particular contribution made by microwave spectroscopy is that, because of its high resolution, many of these interactions are revealed even though their magnitudes are small. For example, magnetic interactions between the spins of electrons and nuclei are, with very few exceptions, much too small to be studied by optical spectroscopy, but readily measured with microwave techniques. The possibility that we shall soon be able to calculate *ab initio* the physical properties of small molecules more accurately than we can measure them is a challenge to the microwave spectroscopist. It is, however, more than the accurate measurement of molecular constants which is at stake. High resolution spectroscopy should enable one to understand the physics of molecules with more insight, and to recognize molecular properties not previously considered.

All this is largely physics, and we turn now to applications in chemistry, particularly the reactive interactions between molecules. There are two distinct levels at which microwave spectroscopy can be used in the study of chemical reactions. The first is as a means of identifying and monitoring the concentrations of particular free radicals; this essentially analytical application has some importance, but it does not make use of the unique features of microwave spectroscopy. These are the ability to monitor rotational and vibrational energy distribution; in conjunction with molecular beam methods (where one can also monitor translational kinetic energy), there is surely a major contribution to be made to our understanding of energy transfer in chemical reactions.

Microwave spectroscopy also has a vital contribution to make to our knowledge of the physics and chemistry of interstellar space. The laboratory microwave study of the OH radical in 1958 can be said to have established the now rapidly developing area of molecular radio and microwave astronomy, and there is an urgent need for microwave data on other free radical species so that their presence in interstellar gas clouds can be sought. Radioastronomical observations have revealed some remarkable dynamical situations in interstellar space (interstellar OH, for example, exhibits maser emission) which can probably only be understood completely with the aid of laboratory microwave studies.

The remainder of this book is organized as follows. Chapter II describes the experimental aspects in detail, whilst Chapter III reviews the theory of the rotational energy levels and their fine and hyperfine structure. Chapters IV and V deal with diatomic and triatomic molecules respectively. Chapter V actually refers to very few molecules, for NCO and HCO are the only short-lived open-shell triatomic species to have been detected and studied by microwave techniques at the present time. Furthermore, molecular ions scarcely appear anywhere in this book, since only one species (H_2^+) has been studied by radiofrequency or microwave techniques. There is plenty of scope, therefore, for the reader who chooses to venture into this field!

II. Experimental Methods

II.1. Microwave Rotational Spectroscopy

II.1.1. Introduction

A microwave absorption spectrometer always contains at least three basic components, whatever other refinements it might possess; these are a microwave generator, an absorption cell containing the gaseous sample, and a detector which monitors the power level after the microwave radiation has traversed the absorption cell. These basic features are shown in Fig. II.1 and

Fig. II.1. Basic components of a microwave absorption spectrometer.

before describing the details it is worth recalling the features common to any microwave absorption experiment. It is usual to sweep the source over a selected frequency range and to monitor the microwave power level at the detector. When the source frequency matches an appropriate transition frequency of the molecules in the absorption cell, the power level at the detector decreases. The sources used in microwave spectroscopy provide monochromatic radiation; often it is necessary to stabilize the source frequency, but at the same time to allow for smooth linear sweeping over an adequate range. The detector is usually a crystal which produces a current proportional to the microwave power level. Many different types of absorption cell are used but, other things being equal, they are designed to maximize the amount of power absorbed by the sample. In most spectrometer systems the absorption is modulated in some way so that the output signal from the detector is an alternating current, which is easier to deal with than a d.c. signal. Other

24

components are added which make it possible to control the amplitude and phase of the microwaves, and to ensure that the detector is working under conditions favouring an optimum signal to noise ratio.

We can now describe in some detail the components of a microwave spectrometer but emphasize that full technical descriptions and the underlying theory must be sought elsewhere. The books by Townes and Schawlow (1955) and Poole (1967) are particularly informative.

II.1.2. Microwave generators

A range of different microwave frequency generators has been developed, but for spectroscopic purposes the most important are the klystron and the backward-wave oscillator (BWO). Both devices make use of the fact that accelerating electrons emit radiation and Fig. II.2 shows, in simplified form,

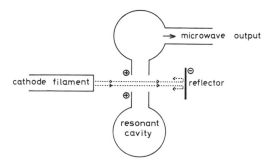

Fig. II.2. Principles of a reflex klystron.

how this effect is employed in the reflex klystron. The cathode is heated electrically so that it produces a continuous beam of electrons by thermionic emission. The electrons are accelerated towards the positively charged cavity, they pass through holes in the cavity walls, and then approach the negatively charged reflector which decelerates them and turns them back towards the cathode. The cavity itself is a small metallic box which stores microwave energy of a particular wavelength, depending on the cavity dimensions in a manner we shall discuss in section II.2. Consequently as the electrons pass through the cavity they are subjected to an oscillating electric field, the phase of which determines whether the electron velocities are increased or decreased. The result of this velocity modulation by the microwave field is to "bunch" the electrons and if the microwave phase is such as to slow a bunch of electrons as they enter the cavity, the cavity receives energy and continues to support the microwave field. The microwave radiation in the cavity can be tapped off through an appropriate window, as shown in the diagram. For a particular cavity size, supporting a particular microwave frequency, the reflector voltage must be adjusted so that the electron velocities satisfy the phase requirements described above. In practice there are actually several different reflector voltage ranges which support separate "modes" for which the klystron will oscillate. Figure II.3 shows a typical example of the frequency and power output of a reflex klystron as a function of reflector voltage.

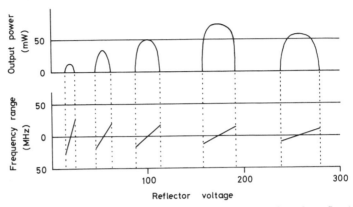

Fig. II.3. Frequency and power output in the oscillating modes of a reflex klystron.

Within a particular mode the frequency is electrically tunable over a range of 50–100 MHz; larger frequency changes (up to several GHz) can be made by changing the size of the cavity mechanically. Hence in spectroscopic applications the cavity is usually adjusted mechanically to a desired centre frequency, and is then frequency-swept by linear variation of the reflector voltage within the range of oscillation for a chosen mode.

The microwave output of a reflex klystron (typically 50–100 mW) powered by well stabilized supplies is essentially monochromatic and reasonably stable in frequency and amplitude. Nevertheless for very high resolution work it is usually necessary to provide additional stabilization, and some methods of effecting this are described later.

The backward-wave oscillator (BWO), which is now being used increasingly as the source for microwave spectroscopy, is illustrated in Fig. II.4. An electron beam is produced by thermionic emission from the cathode; it is collimated by the grid and anode, accelerated by the positively charged accelerator, focussed by a permanent magnetic field, and collected at the positively charged collector plate. The focussed beam passes through a wire helix (held at the same potential as the accelerator) which is a microwave transmission line equal in length to several wavelengths of the lowest required output frequency. When the BWO is turned on and the electron beam produced, random noise in the beam induces voltages in the helix, which in turn velocity modulate the beam and produce bunches of electrons which move towards the collector. As the electron bunches pass the spaces between the helix turns, the electric fields they produce appear outside the helix. When the

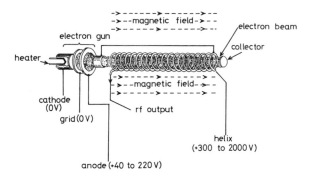

Fig. II.4. Principles of a backward-wave oscillator (based on a drawing in the Hewlett-Packard manual describing their 8690B sweep oscillator).

correct accelerating potential is used, these electric fields are synchronous with the electron bunches along the helix and a backward moving wave is generated in the helix. This wave further bunches the beam, which in turn amplifies the backward wave, until maximum bunching and maximum amplitude of the backward wave are produced. The microwave signal generated on the helix is coupled out of the tube via a d.c. blocking capacitor. The oscillation frequency is determined by the potentials of the accelerating electrode, the helix and the collector (which is often internally connected to the helix).

Compared with the klystron, the backward-wave oscillator has the great advantage of being electronically tunable over its entire range of oscillation, usually several GHz. Although its frequency stability is not intrinsically as high as that of a klystron, modern methods of frequency stabilization make it competitive for many spectroscopic applications.

II.1.3. Microwave propagation, control and detection.

(a) *Waveguide propagation*
Low frequency and radiofrequency electromagnetic radiation is propagated along coaxial cable, but at higher frequencies the attenuation losses increase and propagation through waveguide is generally preferred, certainly from 3 GHz and higher. The waveguide used in microwave spectroscopy nearly always consists of hollow rectangular metal tubing. Different types of wave pattern (modes) can be supported in rectangular waveguide, but the dimen-

sions of the guide impose a restriction on the longest wavelength which can be propagated. This wavelength, called the cut-off wavelength (λ_c) is given by the expression

$$\frac{1}{\lambda_c^2} = \frac{m^2}{(2a)^2} + \frac{n^2}{(2b)^2} \tag{II.1}$$

where a and b are the guide dimensions as shown in Fig. II.5 and m and n are integers which define the wave mode. If $a > b$, the longest wavelength mode

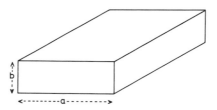

Fig. II.5. Guide dimensions in rectangular waveguide.

is that for which $m = 1$, $n = 0$ and the mode is called a transverse electric (TE_{mn}) mode, in this case TE_{10}. The term "transverse electric" arises from the fact that the oscillating electric field in the guide is perpendicular to the long axis of the waveguide. Figure II.6 illustrates the direction and amplitude of the microwave electric (E) and magnetic (B) fields within the guide, and also the flow of currents induced in the walls of the guide by the microwave fields. The TE_{10} mode is particularly good for spectroscopic purposes since with the correct choice of a and b it is the only mode propagated; furthermore it is possible to cut a longitudinal slot in the centre of the broad face because the current flow in the walls is not disrupted. We shall see that these important properties are utilized in spectroscopic applications. Of course, the metal walls have a finite electrical resistance and even if low resistance metals like copper and silver are used, the microwave power is attenuated during propagation. This attenuation is not too serious at centimetre wavelengths, but increases rapidly at shorter wavelengths.

Circular waveguide can be used for microwave propagation but rectangular is nearly always preferred. However when we come to consider microwave cavities, we shall see that circular and rectangular resonators are both important.

(b) Circuit components
Later in this chapter we shall discuss a number of the microwave circuits used in different spectroscopic applications, and we now briefly describe the main circuit components which are used; they are illustrated in Figs II.7 and II.8.

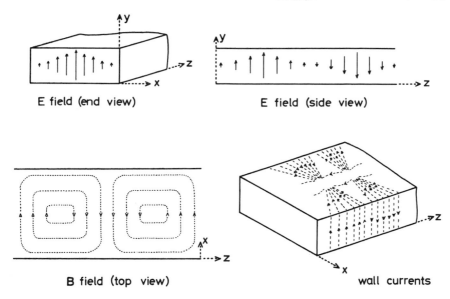

Fig. II.6. Direction and amplitude of the microwave electric (E) and magnetic (B) fields for the TE_{10} mode in rectangular waveguide.

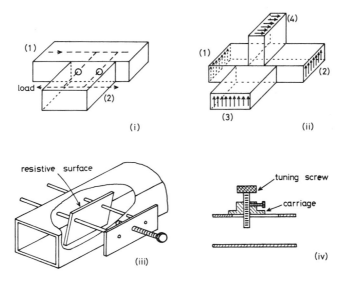

Fig. II.7. (i) Two-hole directional coupler. (ii) Magic T. (iii) Vane attenuator. (iv) Slide-screw tuner.

(i) *Directional coupler*. It is often necessary to divide the microwave power between two different arms and this is accomplished by means of a directional coupler. Figure II.7(i) illustrates a typical two-hole directional coupler. If the wavelength of the radiation in the guide is λ_g, the two holes which are placed in either the narrow or the broad face of the guide must be $\lambda_g/4$ apart. A wave travelling from left to right in the initial guide (1) is split at the holes and continues in the same direction in the side-arm (2); the device can be made unidirectional by absorbing any reflected wave travelling in the side-arm from right to left with an appropriate load of high dielectric material, as indicated. The proportion of the initial power coupled into the side-arm may be increased by increasing the size and number of the holes. The proportions coupled range in most spectroscopic applications from 0·1% (a 30 dB coupler) to 50% (a 3 dB coupler).

An important variant of the directional coupler is the cross-guide coupler (not illustrated) which consists of two waveguides coupled perpendicular to each other, with all four ends accessible. If one end is closed with an absorbing load, this is equivalent to the directional coupler described above.

(ii) *Magic T*. Another device for dividing microwave power is the magic T, illustrated in Fig. II.7(ii). It possesses a plane of symmetry as shown, and a wave entering arm (3) which has its electric vector symmetric with respect to the symmetry plane, cannot directly couple into arm (4). Rather, it will divide equally into arms (1) and (2); however, waves from arms (1) or (2) can couple into arms (3) and (4).

(iii) *Attenuator*. The purpose of an attenuator in a waveguide propagation line is to absorb some of the microwave power and hence to lower the overall power level. The absorbing element is usually a sheet of high dielectric loss material, which is either lowered into the centre of the waveguide through a longitudinal slot in the broad face, or pushed in from the narrow face of the waveguide, where the microwave E field is very small, towards the centre where the E field is at a maximum. The second type, known as a "vane" attenuator, is illustrated in Fig. II.7(iii). The amount of attenuation is controlled by varying the position of the absorbing vane.

(iv) *Phase shifter*. The wavelength of the radiation in a waveguide (λ_g) is not the same as the free space wavelength (λ) but depends upon the wide dimension (a) according to the equation

$$\lambda_g = \lambda/[1 - (\lambda/2a)^2]^{\frac{1}{2}}. \tag{II.2}$$

If the dimension a is changed in some way, the result is the same as if the effective length of the waveguide is changed, and thus the phase of the wave pattern in the waveguide system is altered. In a typical phase shifter the dimension a is effectively reduced by introducing a low-loss dielectric plate parallel to the narrow face of the waveguide, and making provision for this plate to be moved towards the centre of the guide.

(v) *Slide-screw tuner*. A screw tuner is a device in which a metal screw is inserted into the waveguide line; it reflects the microwaves, thereby altering the impedance of the transmission line. Various types of tuner are used but in Fig. II.7(iv) we illustrate the slide-screw tuner which is particularly important in spectroscopic applications. The penetration depth of the screw can be adjusted, and its longitudinal position can be changed. To a first approximation the magnitude of the reflection depends upon the depth of insertion, whilst the phase of the reflection depends upon its position. It is used to match the impedance of the transmission line to that of a device at the end of the line, such as the coupling hole with a microwave cavity.

(vi) *Ferrite devices*. Ferrite materials have the important properties of low electrical conductivity and high magnetic permeability. The low conductivity means that microwave propagation can be accomplished without appreciable attenuation; however, the high magnetic permeability is associated with the presence of electrons with unpaired spins, which can be oriented by an applied magnetic field. If the ferrite is placed in a circular waveguide and the electron spins are aligned by a magnetic field along the direction of propagation, the microwaves interact with the electron spins in such a way that their plane of polarization is rotated. This rotation, called the Faraday rotation, has the important property of being anti-reciprocal. If a wave moving in one direction is rotated through an angle θ, a wave moving in the opposite direction is rotated a further angle θ. An important device which utilizes this property, called an isolator, is illustrated in Fig. II.8(i). The input and output guides are oriented at 45° with respect to each other and the strength of the magnetic field is adjusted to give a Faraday rotation of 45°. Hence a wave travelling from left to right is transmitted by the isolator, but a reflected wave travelling right to left is rotated 90° with respect to the incoming wave, is therefore out of phase with the waveguide, and cannot propagate. The effect of the isolator is thus to permit microwave propagation in one direction only, a property which is important in many microwave circuits. For convenience the isolator usually also incorporates a waveguide twist so that the input and output waveguides have the same relative orientation.

The *circulator* is another device which utilizes the properties of ferrites

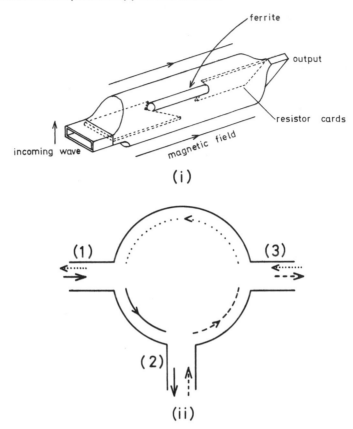

Fig. II.8. (i) Ferrite isolator. (ii) Three-port circulator.

described above. We shall not go into details, but simply describe the properties of a three-port circulator, illustrated in Fig. II.8(ii). Microwaves entering arm (1) of the circulator emerge from arm (2), but do not appear at arm (3). However power reflected from arm (2) emerges at arm (3) and not at arm (1). These properties make the circulator a central component in many spectrometer systems, as we shall see.

(c) *Microwave power detection*
In a microwave spectroscopic experiment, the aim is to detect the emission or absorption of microwave radiation resulting from transitions between molecular energy levels. Hence an important component of the spectrometer system is a detector which permits continuous monitoring of the power level

after the microwave radiation has passed through the gas sample cell. The function of the detector is to convert the microwave energy to a low frequency current, and although several types of detector are available, we here discuss only crystal detectors since these are at present almost always used in spectroscopic applications.

A crystal detector consists of a very fine metal whisker (often tungsten) in point contact with a semiconductor (often a suitably doped silicon crystal). The contact resistance is greater in one direction than in the other and since there is also a very small contact capacitance, the crystal acts as a rectifier which is sensitive to radiation at microwave frequencies. The crystal is inserted into the waveguide line and the incident power causes a voltage drop across the crystal so that, with a suitable circuit, a current (I) flows. This current can be amplified and detected by normal electronic techniques. If the incident microwave power level is in the microwatt range, or lower, the rectified current is directly proportional to the incident microwave power, and since this is in turn proportional to the square of the voltage drop across the crystal, the detector is known as a square-law detector. At higher incident microwave power levels the rectified current is proportional to the first power of the voltage and the detector is then a linear detector (see Fig. II.9).

The usable sensitivity of the detector is, of course, determined by its noise power output, which must therefore be reduced as much as possible. For reasons which will become clear below, it is desirable to modulate the microwave power incident on the crystal so that the output current is an alternating current rather than a direct current. The efficiency with which a small input signal is converted to a change in the output current thus increases with increasing current in the square law region up to a maximum constant value in the linear region. For this reason, it is common practice to "bias" the crystal to ensure that it is operating in the linear region.

Two contributions to the noise power are important, the thermal or "Johnson" noise which depends upon the temperature, and the crystal conversion noise which is proportional to the square of the crystal current. The total noise output power (P) from these two sources is given by

$$P = \left(kT + \frac{CI^2}{v}\right)\Delta v \qquad (\text{II.3})$$

where C is a constant for a particular crystal (typically about 10^{-7} ohms), v is the output frequency (the modulation frequency) and Δv is the bandwidth of the detector system. Clearly, the noise will be minimized by working at a high enough frequency v to render the second term in (II.3) negligible, and by reducing the detector bandwidth Δv. It should be noted, however, that even if

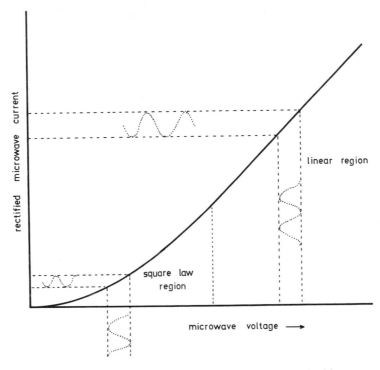

Fig. II.9. Rectified current as a function of microwave voltage incident on a crystal detector.

the modulation frequency v is high, the second term in equation (II.3) becomes significant if the crystal current is too large. Hence there is an optimum bias level at the detector, defined by poor signal conversion efficiency of the crystal at the low power end, and by the emergence of crystal conversion noise at the high power end.

Two methods of modulating the amplitude of the incident power are important in microwave spectroscopy. The first is to modulate the microwave absorption by perturbing the energy levels involved in the spectroscopic transition in a suitable periodic manner (for example, with an oscillating electric or magnetic field). The second method is to mix the microwave power of frequency v_s at the crystal with a second slightly different microwave frequency v_l. The power incident at the crystal is then modulated in amplitude at the "beat" frequency, equal to the difference, $v_s - v_l$, of the two microwave frequencies. A typical and convenient "beat" frequency is 30 MHz, for which the crystal noise term in (II.3) is essentially negligible. This "beat" method is called superheterodyne detection; in practice it is usually combined with molecular modulation, the advantage being that the overall bandwidth of the detection system can then readily be made very small.

II.1.4. Molecular modulation and signal detection

We now discuss details of the techniques used for modulating the power at the crystal detector, particularly molecular modulation methods. Since most molecular energy levels are sensitive to applied electric or magnetic fields, it follows that modulation of the energy levels involved in a spectroscopic transition by oscillating fields will lead to modulation of the microwave power absorbed.

We start by considering the simple two-level system illustrated in Fig. II.10(a). In the absence of any external fields the transitions between E_1 and E_2 induced by microwave radiation will result in net absorption, and if the spectrum is obtained by slowly sweeping the microwave frequency through the transition frequency v_0, the detector will register a decrease in microwave power and hence a change in d.c. output. This simple method is called crystal video detection. Suppose now that a static electric field is applied and produces a simple Stark effect as illustrated in Fig. II.10(b). On plotting the absorption spectrum we now find that the absorption at v_0 is replaced by absorption at a higher frequency v_1, the shift in the line being dependent on the magnitude of the applied static voltage V. This shift is called a Stark shift.

Now suppose that instead of a static electric field, we apply an oscillating field by feeding a square-wave voltage to an electrode placed in the gas sample; suppose also that the square-wave is zero-based, that is, the voltage changes

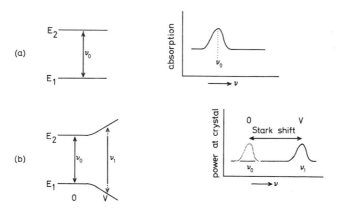

Fig. II.10. Stark effect and line shift for a two-level system.

from 0 to V. Figure II.11(a) shows the microwave power level at the crystal as the microwave frequency v is swept slowly through the same range as in Fig. II.10(b). When the square-wave voltage is in the first half of its cycle (0 volts), absorption centred at v_0 is obtained, but in the second half of each cycle (V volts), the Stark component centred at v_1 is recorded. The output from the crystal detector is now an alternating current, and after amplification with an amplifier tuned to the modulation frequency, the signal output is as shown in Fig. II.11(b). The unshifted line at v_0 is represented by an oscillating output in which the peaks correspond to the bottom of the square-wave, that is, it is 180° out-of-phase with the square-wave modulation. On the other hand, the Stark component at v_1 is also represented by an oscillating signal, but the peaks now correspond to the top of the square-wave (V volts); in other words the output signal is in-phase with the modulation. Instead of displaying these tuned amplifier outputs, it is usual to feed them to a phase-sensitive detector. This is an electronic circuit in which the signal is compared with the original square-wave modulation, producing a d.c. output which is proportional to the cosine of the phase difference between them. In this case the phase difference is either 180° or 0°; the d.c. output is thus proportional to −1 or +1 and hence display of the d.c. output yields the spectrum shown in Fig. II.11(c). Note that the Stark component is inverted with respect to the unshifted line.

Several important points of detail should now be noted. First, we have achieved the aim of providing the crystal detector with a modulated microwave power level and in order to minimize the crystal conversion noise in equation (II.3), we choose as high a modulation frequency as is convenient. In practice a modulation frequency of 100 kHz is often employed. Second, for purposes of illustration we have assumed the amplitude of the square-wave modulation to be large enough to completely separate the absorption line and its Stark component. As the modulation amplitude is decreased, the separation between the lines decreases and eventually they overlap. The resulting line shape is then essentially that of a first derivative of the original absorption line in Fig. II.10(a). Third, we have chosen a particularly simple example in which the applied electric field merely shifts the energy levels; usually the Stark effect is more complicated, in that the levels actually split into two or more components. There are then several Stark components and since half of the total absorption intensity is divided between them, whilst the other half remains in the unshifted line, the Stark components in complicated spectra can become too weak to be observed.

The Stark square-wave modulation is normally zero-based but it is important to consider the case when the modulation is from −V volts to +V volts. The Stark effect on the energy levels is not sensitive to the sign of the

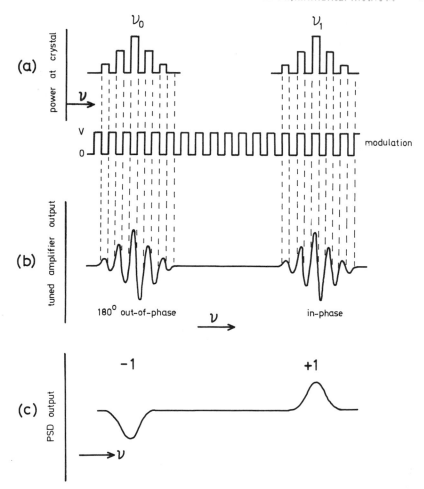

Fig. II.11. Principles of Stark modulation and phase-sensitive detection.

Stark voltage, so that provided the rise time of the square-wave is negligible, only the Stark component will be observed. However the effect of the Stark modulating field is exactly the same as that of a static field, so that the power at the crystal is not modulated. If we use sinusoidal electric modulation, rather than square-wave, it is again desirable to provide an additional static bias field so that the sine-wave oscillation voltage is from 0 to V volts, rather than from $-V$ to $+V$. Sine-wave modulation has some disadvantages, however, as reconsideration of Fig. II.10 will indicate. Since the voltage in sine-wave modulation varies continuously, we no longer have the simple case of an unshifted line and a separated Stark component. Absorption over the whole range of frequencies from v_0 to v_1 will occur, and for this reason if sine-wave modulation must be used, its amplitude must be small enough to keep the Stark shift within the width of the absorption line. If the amplitude of the modulation is increased the line is eventually broadened beyond detection, rather than yielding two sharp components as in the case of square-wave modulation.

In the study of open-shell molecules the presence of large magnetic moments means that the molecular energy levels are sensitive to applied magnetic fields, which can therefore be used for modulation purposes. It is, however, difficult to produce a square-wave magnetic field (Zeeman) modulation, and the modulation is therefore usually sinusoidal and suffers from the disadvantages outlined above. With magnetic field modulation, therefore, it is desirable to provide a static bias magnetic field (so that the modulation is zero-based) and essential to avoid line broadening due to too large a modulation amplitude. We shall see that magnetic field modulation is frequently used in gas phase electron resonance experiments. The actual methods used for applying the modulation fields, electric or magnetic, will be described in section II.1.6.

Apart from improving the signal-to-noise ratio by suppressing the crystal conversion noise, high frequency modulation and phase-sensitive detection improves the sensitivity in other ways. The use of a tuned amplifier means that noise components at other frequencies are not amplified. Furthermore the time constant of the output stage of the phase-sensitive detector can be made as long as 1–100 seconds, so that all but the lowest frequency noise components are filtered out.

Before leaving the subject of modulation we should mention other techniques which can be used. It is possible, for example, to modulate the frequency of the microwave source which will, in principle, give a similar result to zero-based molecular modulation by applied fields. Source modulation can create other problems, however, and is usually avoided if possible. In molecular beam studies mechanical chopping of the beam also has the effect of modulating the power amplitude at the detector, and is sometimes more convenient to apply than field or frequency modulation.

II.1.5. Complete microwave rotational spectrometer systems

We are now in a position to discuss the complete spectrometer system since we have dealt with all the essential components, apart from the gas absorption cell. We will discuss the various types of cell which can be used in section 1.6; for the moment it suffices to say that the absorption cell is filled with gas at a desired pressure, the microwave radiation is propagated through the cell, and some means of applying electric or magnetic molecular modulation is provided.

It is probably true to say that every home built microwave spectrometer is different from its predecessors; assembling microwave circuits yields much the same sort of satisfaction as does Meccano (i.e. Erector set) or Lego. Consequently we shall here describe two different spectrometer systems which are similar to others in active use. Our concern is to illustrate how the principles described in the previous sections are applied in practice.

Figure II.12 illustrates a simple modulation spectrometer. The source is usually a klystron or a backward-wave oscillator, either stabilized by being phase-locked to a harmonic of a stable reference oscillator, or free-running if very high frequency stability is not required. The microwave radiation passes through an isolator (which prevents reflected radiation from upsetting the source), a directional coupler which taps off a small sample for frequency measurement, a variable attenuator which enables the power level in the gas sample cell to be adjusted, and finally a crystal detector. The modulator provides an oscillating electric or magnetic field in the cell, and the consequent a.c. signal from the crystal detector is amplified and fed to a phase-sensitive detector, where it is compared with a reference signal also provided by the modulator. The d.c. signal output from the phase-sensitive detector is fed to the Y input of an $X–Y$ recorder. Microwave frequency sweep (typically at rates between 1 kHz per second and 1 MHz per second) is provided by a sweep voltage generator, which also provides the X input to the recorder. Consequently the microwave absorption spectrum is presented directly as a function of microwave frequency.

Figure II.13 describes an important refinement of the first spectrometer. It might be recalled that in our earlier discussion of crystal detectors, we pointed out the desirability of presenting the crystal with an optimum total power level. From Fig. II.12 it will be clear that the total power reaching the crystal is proportional to the power level in the absorption cell. In order to maximize the absorption by the sample gas, we want to use the highest possible microwave power level consistent with avoiding saturation effects. The spectrometer shown in Fig. II.13 allows one to vary the total power level at the crystal, whilst also permitting independent variation of the power level in the sample cell. This is achieved by dividing the total power from the source at the second directional coupler. Part of the power passes via a variable attenuator through the sample; the other part passes through a balancing arm containing a phase shifter and a variable attenuator. The two waveguide lines join up again at the third directional coupler and hence the power level at the crystal can be altered by changing the level in the balancing arm; at the same time the

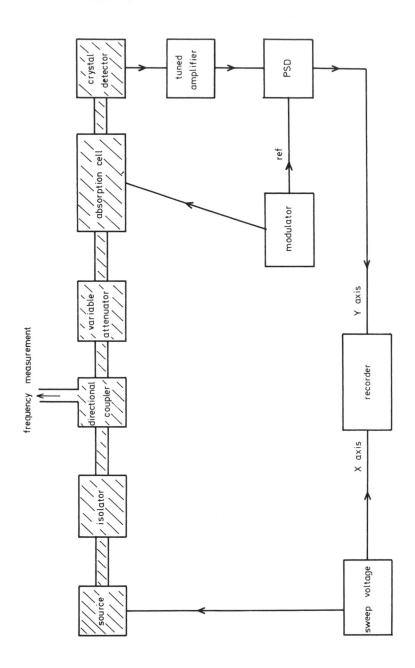

Fig. II.12. Block diagram of a microwave absorption spectrometer employing molecular modulation.

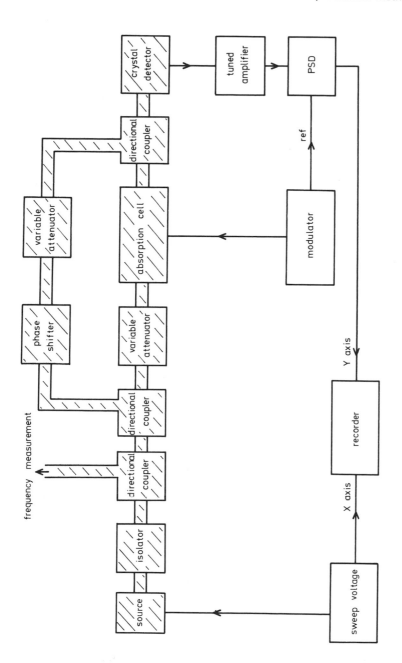

Fig. II.13. Block diagram of a microwave absorption spectrometer with a balancing side-arm to provide independent variation of the bias power at the detector crystal.

power level in the cell can be optimized independently. Another advantage is that undesirable reflections in the absorption arm can be compensated for by adjustment of the balancing arm; this is particularly important when accurate intensity measurements are required.

We conclude by mentioning the tunable-cavity spectrometer, which has been particularly important in free radical studies. Since this is actually similar in design and operation to an electron resonance spectrometer, we defer our description until section II.2.

II.1.6. Absorption cells and free radical studies

We now come to the most important part of a microwave spectrometer, namely, the absorption cell. Everything discussed so far in this chapter is equally relevant to both closed and open-shell molecules. However in the study of short-lived free radicals the main problem is to establish and maintain a sufficiently high concentration inside the absorption cell; for this reason, the design of the cell is crucial.

In the microwave rotational spectroscopy of stable molecules, the absorption cell is nearly always a length of the appropriate rectangular waveguide, sealed at both ends with a suitable material (for example, mica or teflon) which permits penetration of the microwave radiation whilst also providing a good vacuum seal so that the pressure in the cell can be lowered. For given cross-sectional dimensions, rectangular waveguide will propagate radiation of any wavelength shorter than the cut-off wavelength, but the attenuation increases at higher frequencies. The optimum length of a waveguide absorption cell is inversely proportional to α_0, a measure of the attenuation due to the finite electrical resistance of the walls. At centimetre wavelengths this optimum length turns out to be a few metres, and this is the first practical difficulty confronting the would-be free radical spectroscopist. In order to make use of the inherent sensitivity of the spectrometer, the absorption cell must be filled with free radical species, a practical impossibility if the cell is several metres long and the radical is short-lived. Hence one must compromise and choose a cell as long as possible, but one which nevertheless enables a reasonable filling factor for free radicals to be achieved.

Another important factor in the design of an absorption cell is the method of introducing field modulation, particularly Stark modulation. With rectangular waveguide cells, two methods are common. In the first, shown in Fig. II.14(a), a thin metal septum is inserted into the centre of the guide, parallel to the broad faces. The septum is supported at the walls with a suitable insulator and the Stark voltage applied to the septum, with the waveguide walls at earth potential. In this orientation the oscillating Stark field is parallel to the microwave electric field. In the second method (Fig. II.14(b)) the waveguide is cut into separate halves which are held in place by insulators, and the Stark modulation is applied between the two halves.

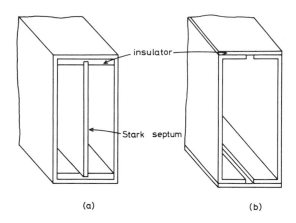

Fig. II.14. Two methods of introducing Stark modulation in a rectangular waveguide absorption cell: (a) Stark septum; (b) split waveguide.

The free radicals are usually produced either by passing a gas through a radiofrequency or microwave discharge, or by reacting the products of a discharge with a secondary gas. Continuous flow methods are used and the discharge is situated close to the absorption cell. Charged species present in the discharge plasma give rise to unacceptable attenuation of the microwave radiation, and for this reason the discharge zone must remain outside the absorption cell. Molecular free radicals are usually very short-lived at pressures above 0·1 torr and the simple discharge method has been successful in very few cases. However atomic species produced by an appropriate discharge tend to be longer lived and they can be successfully pumped into the absorption cell; mixing the atoms with a secondary gas inside the cell has proved to be a fairly successful technique for generating detectable concentrations of free radicals.

An early example is the successful study of the SO radical by Powell and Lide (1964). They used a cell similar to that described in Fig. II.14(b) and produced the SO radical by reacting O atoms (produced by flowing molecular oxygen through a radiofrequency discharge) with either sulphur vapour or carbonyl sulphide. The reaction

$$O + OCS \longrightarrow SO + CO \tag{II.4}$$

leads to a fairly high concentration of SO radicals which are relatively long-lived. Powell and Lide's absorption cell is illustrated in Fig. II.15; they used Stark modulation at 80 kHz and were able to measure absorption lines ranging from 13 to 66 GHz. Independent work by Amano, Hirota and Morino (1967) using a similar absorption cell yielded the same results.

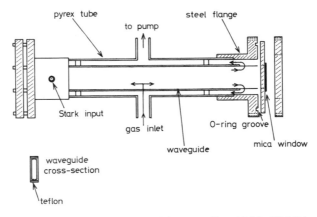

Fig. II.15. Waveguide absorption cell used by Powell and Lide (1964) in the study of the SO radical.

We mentioned in section II.1.3 that microwave radiation can be propagated along cylindrical waveguide and the first observation of a free radical microwave spectrum was achieved by Dousmanis, Sanders and Townes (1955) who used a cylindrical cell to detect OH radicals in the products of a water vapour discharge. Figure II.16 shows the absorption cell used by them. It consisted of a cylindrical brass tube (approximately 150 cm long and 3 cm in diameter), coupled at both ends to standard rectangular waveguide through tapered sections. Inside the brass tube was a quartz tube with inlet and outlet ports for gas flow. Water vapour was pumped continuously through a radiofrequency discharge, into the quartz cell, and out to a cold trap. High frequency magnetic field modulation was provided by passing an alternating current through a wire coil wound around the outside of the brass cylinder. Since a high frequency magnetic field will not penetrate a normal brass wall, longitudinal slots were cut along the length of the tube; a d.c. bias magnetic field was also applied. The modulating field inside the tube was small and the microwave attenuation by the cell was large, but nevertheless absorption lines at frequencies ranging from 7·7 to 37 GHz were observed and measured. In retrospect, this work is a landmark in spectroscopy; not only did it give needed hope and stimulus to others searching for free radical spectra, it also led to the

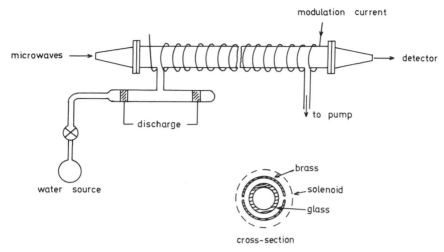

Fig. II.16. Cylindrical cell used by Dousmanis, Sanders and Townes (1955) in the study of the OH radical.

detection of interstellar OH and thereby to the establishment of molecular microwave astronomy.

So much for waveguide cells, which are not now used much for free radical spectroscopy. The SO studies were successful because SO is fairly long-lived, and the OH experiments succeeded primarily because OH in water vapour regenerates itself for a considerable distance beyond the discharge zone. Apart from the chemical difficulties, however, waveguide cells become insensitive at higher frequencies and there are better alternatives. Kewley, Sastry, Winnewisser and Gordy (1963) have made measurements up to 250 GHz on the free radicals SO and CS using a so-called "free space" absorption cell (note in passing that CS is long-lived and has a closed-shell ground state). The cell and discharge system are illustrated in Fig. II.17. The cell of length 40 cm and diameter 10·5 cm was made of pyrex and was connected to three separate discharge tubes where the SO and CS radicals were produced by discharges in SO_2 and CS_2 respectively. The ends of the cell were sealed by teflon windows which do not attenuate the microwave radiation appreciably. The radiation emerges from a radiating horn, is focussed by a teflon lens to produce a collimated microwave beam passing through the cell, and is collected by a second lens and horn. Hence the absorption cell is simply placed in the path of the microwave beam, and it is for this reason that the cell is called a free space cell. Clearly the size and geometry of this arrangement make it difficult to apply modulating fields, and the spectrometer therefore used simple crystal video detection. The inherently low sensitivity of such

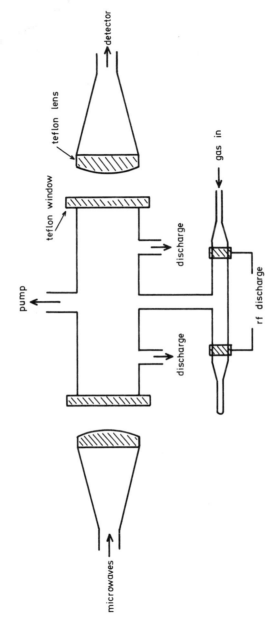

Fig. II.17. Free space cell used by Kewley, Sastry, Winnewisser and Gordy (1963) in the study of the SO and CS radicals.

an arrangement is, however, compensated for by the more intense absorption at the higher frequencies. The reader will recall from Chapter I that the net power absorbed depends on the population difference between the levels, the quantum of energy absorbed, and the Einstein B coefficient. The first two of these quantities increase with increasing frequency, allowing one to work with simpler detection systems in favourable cases. Nevertheless higher sensitivity is needed because not many free radicals are as long-lived as SO and CS.

Most free radical microwave spectroscopists are currently using the "parallel plate" cell, an example of which is illustrated in Fig. II.18. The microwave radiation is propagated and collected by a pair of suitable horns, and is transmitted between the parallel metal plates, usually in a TE mode. The plates themselves are enclosed in a suitable vacuum chamber through which the gas is continuously pumped. An attractive feature of the parallel plates is that besides transmitting the microwave radiation, they form convenient electrodes for Stark modulating and static fields, the latter providing splittings from which dipole moments can be determined. Saito and Amano (1970) have used such a cell for studying the NCO radical. The plates in their spectrometer are 4 cm wide and 40 cm long, and the gap between them is 4·3 mm. They are held apart by perspex blocks and fed with 100 kHz Stark modulation. The NCO radical was produced by reacting fluorine atoms (formed by passing CF_4 through a microwave discharge) with cyanic acid (HNCO), the mixing occurring just inside the parallel plates. Saito and Amano reported that their cell operates satisfactorily up to at least 130 GHz and the sensitivity of the overall system seems to be high.

The parallel plate cell represents the best current compromise for free radical rotational spectroscopy but there is much room yet for innovation and, as we shall see, the techniques used in electron resonance experiments are more sensitive and also more convenient from a chemical viewpoint.

II.2. Gas Phase Electron Resonance

II.2.1. Introduction

Electron resonance of gaseous molecules has much in common with microwave rotational spectroscopy, and relatively little in common with electron resonance of liquids and solids. The experimental techniques used for gases are similar to those employed in condensed phase studies, but the interpretation is very different. In this chapter we concentrate on experimental techniques, but later in this book when we consider the interpretation of the results, we shall see how closely related and complementary are electron resonance and rotational spectroscopy.

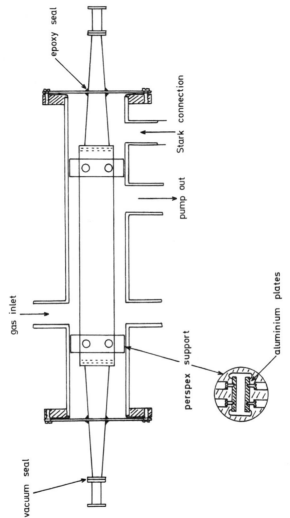

Fig. II.18. Parallel plate cell (based on a diagram kindly supplied by D. R. Johnson, 1973).

The most important feature of the electron resonance experiment is that the spectrometer is designed to operate at a *fixed* microwave frequency and the molecular energy levels are tuned by an applied static magnetic field. In condensed phases the relevant energy levels are fairly accurately described as electron spin levels which, because of the spin magnetic moment, are readily tunable by an applied magnetic field. These spin levels are, indeed, usually degenerate in the absence of the field, which is therefore necessary to establish a separation between the levels before a spectroscopic transition can be induced. Since the transition occurs because of coupling between the oscillating magnetic component of the microwave radiation and the electron spin magnetic moment (i.e. a magnetic dipole transition), the experiment is justifiably described as "electron spin resonance". In contrast, the gas phase experiment is much more general. The energy levels are not necessarily degenerate in the absence of an applied field, and although it is usually the presence of unpaired electron spin which makes it possible to tune the levels magnetically, they cannot, in general, be described as spin levels. Moreover the transitions studied are usually induced by coupling of the molecular electric dipole moment with the electric component of the oscillating field (i.e. they are electric dipole transitions). Consequently we have not included the word "spin" in the title of this section, although many might have expected to see it there.

An electron resonance spectrometer therefore differs from a microwave rotational spectrometer in at least two major respects. The most obvious difference is the presence of a large electromagnet which usually dominates the laboratory housing it and condemns the human occupants to life in the basement. The second difference is less obvious visually, but just as important. Because the microwave frequency is not swept, the absorption cell used is quite different from those described in section II.1.6. Instead one uses a small resonant cavity, designed for a particular fixed frequency, which stores microwave power and which, it turns out, is very suitable for free radical investigations because of its small volume. Our first task, therefore, is to describe the most important features of resonant cavities, and to show how their properties are used to advantage in free radical spectroscopy.

II.2.2. Microwave resonant cavities

A resonant cavity is a hollow space enclosed by metallic walls which can store electromagnetic radiation of certain particular wavelengths. It can be regarded as a multiple reflection device in which the distance between the reflectors (i.e. the walls) is comparable with the wavelength of the radiation or a small integral or half-integral multiple of the wavelength. Compared with a

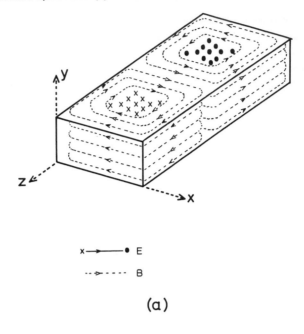

x——→ • E

- -→ - - - B

(a)

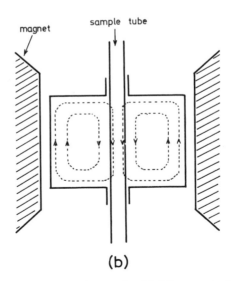

(b)

Fig. II.19. Rectangular TE_{102} cavity mode. (a) Microwave electric and magnetic fields. (b) Orientation of the cavity in an external magnetic field for the detection of magnetic dipole transitions.

waveguide transmission line it is able to store radiation at a high power density, and it is this feature which makes cavity resonators important in spectroscopy. Nearly all of the cavities used in spectrometers are either rectangular or cylindrical, so we will confine our discussion to them.

The modes supported in a rectangular cavity are either TE_{mnp} modes or TM_{mnp}. (TE stands for "transverse electric" and TM for "transverse magnetic.") The subscripts m, n, and p describe the number of half-wavelengths in the standing wave pattern in the x, y, and z directions defined in Fig. II.19(a). The mode shown in the figure is transverse electric and is, in fact, a TE_{102} mode. The wavelength λ_0 of the radiation supported in this mode satisfies the equation

$$\left(\frac{m\lambda_0}{2a}\right)^2 + \left(\frac{n\lambda_0}{2b}\right)^2 + \left(\frac{p\lambda_0}{2d}\right)^2 = 1. \tag{II.5}$$

Strictly speaking this equation only holds if the cavity is completely evacuated. The correction for the dielectric constant of air at atmospheric pressure is negligible for our purposes, but insertion of solid or liquid material usually affects the resonance frequency quite drastically. The TE_{102} mode is commonly used in electron resonance studies, particularly if magnetic dipole transitions are being studied, as is always the case for condensed phase samples, and often the case for gases (such as, for example, molecular oxygen or any atom). The requirements for observation of magnetic dipole transitions are that the sample tube be placed in the region of maximum microwave magnetic field, and the cavity be oriented so that the microwave magnetic field is perpendicular to the direction of the external magnetic field, as shown in Fig. II.19(b).

In a TE_{mnp} mode cylindrical cavity the integer m represents the number of E field maxima (B field for a TM mode) in a 180° angle measured in a plane perpendicular to the axis of the cylinder, n represents the number of E field maxima (B field for a TM mode) between the centre and wall, and p is the number of E field maxima along the axis of the cylinder. The resonant wavelength of a cylindrical cavity (TE or TM mode) is given by

$$\lambda_0 = \frac{2}{\{(2x_{mn}/D)^2 + (p/L)^2\}^{\frac{1}{2}}} \tag{II.6}$$

where D is the diameter, L is the length and x_{mn} is a constant related to a Bessel function (tabulated). Most gas phase electron resonance work has been carried out using the cylindrical TE_{011} mode, illustrated in Fig. II.20. The oscillating electric field is parallel to the cylinder walls, with zero amplitude at the centre of the cylinder and at the walls, and maximum amplitude halfway between the centre and the walls. The oscillating magnetic field is parallel to the axis of the cylinder, except at the ends. It has maximum amplitude at the centre and at the cavity walls. Hence for magnetic dipole transitions the gas sample should be at the centre of the cavity, and the cavity itself should normally be oriented with the external static magnetic field perpendicular to the cylinder axis. For electric dipole transitions, however, the cavity is usually oriented with the external field parallel to the cylinder axis (i.e. perpendicular to the microwave electric field). We shall see, however, that both magnetic and electric dipole transitions can in some instances be induced by the opposite (parallel) orientation of static and oscillating fields. A further important point about the TE_{011} mode is that the induced wall currents do not flow across the junctions between the cylinder and the ends; this means that the end plates can be electrically insulated from the cylinder body, something we shall discuss again in connection with Stark modulation.

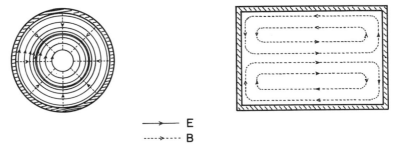

$$\longrightarrow \quad E$$
$$\text{----}\!\!\rightarrow\!\!\text{----} \quad B$$

Fig. II.20. Microwave electric and magnetic fields in a cylindrical TE_{011} cavity.

An important aspect of cavity design is the manner in which the microwave radiation is fed into and out of the resonant cavity. In most cases the cavity is connected to rectangular waveguide, and the radiation passes from one to the other through a small hole or slot called an iris. The coupling is designed to excite preferentially a particular mode by making the electric or magnetic components at the coupling region coincide with those of the desired cavity mode. The size of the iris is critical, but even to the author who has built at least a dozen cavities in his time, iris coupling is a black art achieved either by experiment or by consultation with the right expert!

The final matter for discussion is the figure of merit or Q factor of the cavity. Up to now we have stated that a resonant cavity will, in a given mode, support electromagnetic radiation of a particular frequency, and no other. In practice when we plot the power of the radiation absorbed by the cavity as a function of frequency, we obtain a graph similar to that shown in Fig. II.21 in which the absorption curve has a finite width. The Q factor for the cavity is given by $Q = v_0/\Delta v$ where Δv is the width at half peak height. In practice we attempt to make Q as large as possible, but for each mode there is an upper limit. For the modes we have discussed, Q is usually in the range 5000 to 20,000. In other words, for a TE_{011} mode cavity, resonant at 10 GHz and with a Q of 10,000, the bandwidth Δv is 1 MHz. In a spectroscopic experiment using a resonant cavity, the frequency of the microwave source must be equal or very close (say within 1–10 kHz) to the resonant frequency of the cavity. In the following section we shall describe a common technique by which the source frequency is locked automatically to the cavity frequency. However

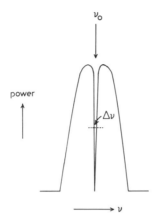

Fig. II.21. Cavity Q curve.

the considerations outlined above show why cavities are used in electron resonance experiments, but not usually in microwave rotational spectroscopy. The fixed frequency operation of an electron resonance spectrometer makes the resonant cavity an ideal sample cell; if, however, one wishes to sweep the source frequency, the cavity must also be made frequency tunable. This can be done, as we shall discuss later, but it is not easy.

II.2.3. Electron resonance spectrometers

Figure II.22 shows a block diagram of a typical electron resonance spectrometer suitable for gas phase work; its main features are as follows. Microwave radiation is generated by the source (a klystron or BWO) and proceeds to arm 1 of the three-port circulator via an isolator, a directional coupler to provide a sample for frequency determination, and a variable attenuator. The power emerges from arm 2 of the circulator and proceeds to the resonant cavity through a slide-screw tuner. Power reflected back from the cavity re-enters arm 2 of the circulator and emerges at arm 3 where it is monitored by

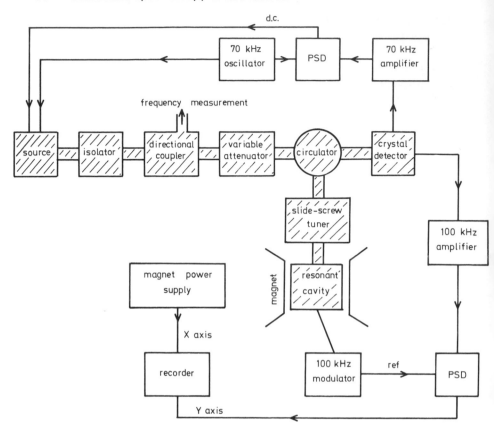

Fig. II.22. Block diagram of an electron resonance spectrometer.

the crystal detector. In setting up the instrument the frequency of the source is locked to the cavity, and the slide-screw tuner and cavity iris adjusted so that little or no power is reflected to the crystal detector. Electron resonance lines are now sought by sweeping the external magnetic field. When a resonance condition is satisfied, power in the cavity is absorbed by the sample, thus changing the impedance of the cavity arm. This causes a change in the reflected power and hence a change in the crystal output current. A modulating electric or magnetic field is applied across the cavity, in addition to the static field, so that as before, the crystal output is an alternating current. The aspects which should now be discussed in more detail are the means of locking the microwave source frequency to the cavity frequency, and the method of applying modulation across the cavity.

An automatic frequency control (AFC) system for locking the source to the cavity is described in Fig. II.23. The frequency of the microwave source is modulated over a small range by applying a 70 kHz voltage to the klystron reflector (or BWO helix, but let us suppose we are using a klystron). If the unmodulated klystron frequency is exactly equal to the cavity resonant frequency (i.e. corresponds to the centre of the cavity Q curve), source modulation will result in a voltage appearing at the crystal detector of frequency

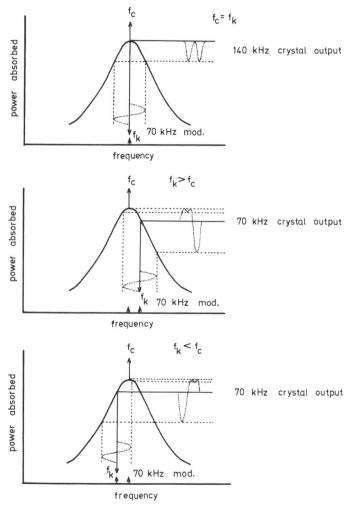

Fig. II.23. Principles of an automatic frequency control (AFC) system for locking the source frequency to the cavity resonant frequency (see text).

140 kHz (but none at 70 kHz), as shown in Fig. II.23(a). If, however, the centre frequency of the klystron is shifted from the centre of the cavity Q curve (Fig. II.23(b)), the voltage appearing at the detector has a large 70 kHz component. However the phase of this 70 kHz detector voltage, relative to the original modulation, depends upon whether the klystron centre frequency is higher or lower than the cavity frequency. Moreover, the amplitude of the 70 kHz signal from the detector increases with the difference in frequency between the klystron and cavity. The 70 kHz voltage from the detector, which is called the error voltage, is amplified in a 70 kHz tuned amplifier and fed to a phase-sensitive detector where it is compared with a reference sample of the original 70 kHz modulation voltage. The phase detector output, a d.c. voltage which depends upon the relative phases and amplitudes of the signal and reference, is then applied to the klystron reflector, thus completing the loop and correcting the klystron frequency back to the cavity frequency. By this means it is not difficult to stabilise the frequency of the klystron to better than 1 part in 10^7. Drifts in frequency will, of course, occur if the cavity frequency changes because of, for example, thermal drift or chemical deposition (which is often a problem in gas phase studies). The AFC system can, however, also be used to lock the source to the cavity whilst slowly sweeping the latter, either by systematically changing its dimensions or by inserting a tuning rod. We will return to this aspect in section II.2.5.

The other aspect of the spectrometer which requires detailed discussion is the field modulation. Everything discussed earlier in section II.1 applies equally to electron resonance. In conventional electron spin resonance of condensed phases the spin levels are sensitive to magnetic fields only and consequently magnetic field modulation is used, 100 kHz being a commonly used frequency. Figure II.24 illustrates a rectangular TE_{102} mode cavity with a pair of modulation coils used to provide an oscillating magnetic field in the same direction as the static field. The walls of the cavity, outside of which the coils are mounted, must be of metal which is thin enough to permit penetration of the modulating field, yet thick enough to stop the microwave power from leaking out of the cavity. Gold or silver plated ceramic walls are often used for this purpose. The cavity shown in Fig. II.24 can be used for magnetic dipole transitions in gases, but actually most gas phase investigations have been concerned with electric dipole transitions and for these the cavity shown in Fig. II.25 has proved to be extremely useful (Carrington, Levy and Miller, 1967a). The cavity resonates in the cylindrical TE_{011} mode and is designed to be gas tight. In other words, the gases under investigation flow directly into and out of the cavity itself. The most novel feature of the cavity, however, is that it is very convenient for Stark modulation. We mentioned in the previous section that the wall currents in the TE_{011} mode do not flow across the junc-

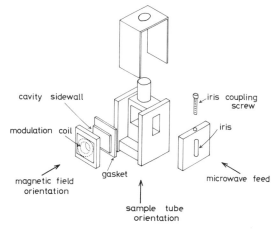

cavity sidewall

modulation coil

magnetic field
orientation

gasket

sample tube
orientation

iris coupling
screw

iris

microwave feed

Fig. II.24. Rectangular TE_{102} cavity (based on a drawing of the Varian Associates V-4545 cavity).

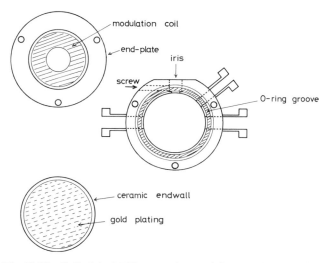

modulation coil

end-plate

iris

screw

O-ring groove

ceramic endwall

gold plating

Fig. II.25. Cylindrical TE_{011} cavity used for gas phase studies.

tions between the cylinder and the end plates. Consequently the end plates can be electrically insulated from the cavity body and used as Stark electrodes. By this means electric field modulation, sine wave or square wave, can readily be applied across the gas sample in the cavity. The cavity is oriented in the magnet gap with the Stark field parallel to the static magnetic field, so that the microwave electric and static magnetic fields are perpendicular. If magnetic field modulation is required, the metal Stark plates are replaced by metal coated ceramic plates with modulation coils mounted outside them.

As in the case of microwave rotational spectroscopy, the field modulation results in the signal from the crystal detector being an a.c. signal at the modulation frequency, which is amplified and phase detected as before. Consequently the electron resonance spectrometer shown in Fig. II.24 contains two separate modulation and phase-sensitive detection systems, one at 70 kHz for the AFC system, and the other at 100 kHz for signal modulation and detection; it is important that these two systems do not interact with each other.

We will discuss the free radical studies themselves in the following section. It is important to realize that although a spectrum is recorded by keeping the frequency fixed and measuring the magnetic fields at which absorption lines occur, it is possible to vary the cavity frequency by changing its dimensions or inserting a tuning rod (often quartz or teflon) and thus recording the spectra at different fixed frequencies within the range of the microwave source. We shall see the importance of this elsewhere.

Many other types of electron resonance spectrometer have been built, covering a wide range of resonance frequencies and detection systems. In particular, superheterodyne detection is often used, especially when work at very low microwave power levels is performed. However we will not go into further details since the superheterodyne detection operates in essentially the same way as described previously in section II.1.

II.2.4. Free radical studies

The first free radical studies using gas phase electron resonance were made by Radford (1961, 1962) who studied the OH radical; his experiments were based on the earlier microwave work of Dousmanis, Sanders and Townes (1955), the products of a microwave discharge in water vapour being passed continuously through the electron resonance cavity. Subsequently Radford (1964) found that he was able to detect the spectrum of SH by adding H_2S to the products of a water vapour discharge at the entrance to the cavity. He was also able to detect the spectra of SH, SeH and TeH by coating the cavity walls with sulphur, selenium and tellurium respectively and passing into the cavity a continuous stream of hydrogen atoms.

The technique of generating free radicals by reactions between atoms and secondary molecules, arranging for the mixing to occur inside the resonant cavity, has been extensively employed by the author and his colleagues. The cavity shown in Fig. II.25 has two entrance ports for the reactant gases and one exit port leading to the pumping system. Among the reactions used to generate diatomic radicals have been the following (the radical species detected are italicized):

$$Br_2 + O \longrightarrow BrO + Br$$

$$CH_3I + O \longrightarrow IO + products$$

$$OCS + F \longrightarrow SF + CO \qquad (II.7)$$

$$OCSe + F \longrightarrow SeF + CO$$

$$OCS + N \longrightarrow NS + CO$$

The technique has also been extended to two linear triatomic radicals through the use of the reactions

$$HNCO + F \longrightarrow HF + NCO$$
$$\qquad\qquad\qquad\qquad\qquad\qquad (II.8)$$
$$HNCS + F \longrightarrow HF + NCS$$

and very recently to a non-linear triatomic radical, HCO, through the reaction of fluorine atoms with formaldehyde. In this last example it is necessary to record the electron resonance spectra at a series of fixed microwave frequencies, for reasons which will be made clear in Chapter V.

The gas phase electron resonance technique has also been applied successfully to certain electronically excited species, namely, O_2, SO, SeO and NF in their excited $^1\Delta$ states. However it should be noted that these $^1\Delta$ states are very stable with respect to both radiative decay and collisional deactivation. Most excited electronic states are destroyed far too rapidly by molecular collisions for conventional microwave absorption techniques to have much chance of success. This is where molecular beam and double resonance methods come into their own, as we shall see.

One final point worth mentioning is that the cavity shown in Fig. II.25 can also be used to subject the gas to a static electric field, in addition to the oscillating field and static magnetic field. The resulting Stark splitting of the levels can be measured and the electric dipole moment of the species determined. We will discuss the details later.

II.2.5. Tunable-cavity microwave spectrometers

We now turn to tunable-cavity instruments, the principles of which really belong to section II.1 although the techniques used are closely related to those employed in electron resonance. We have pointed out that the major problem in free radical studies is in effectively filling the sample cell volume; the relative success of electron resonance methods must be attributed to the use of resonant cavities which are usually of small volume. It follows that the combination of a resonant cavity with swept frequency techniques should prove powerful, provided the experimental problems can be solved. We can change the resonant frequency of a cavity either by changing its dimensions, or by changing the effective dielectric constant of the volume enclosed by the cavity. The latter is usually accomplished by inserting a quartz or teflon tuning rod. The other requirement is to lock the frequency of the source to the resonant frequency of the cavity. We have seen that AFC techniques are well developed in fixed frequency electron resonance spectrometers, and the same methods can be used in a tunable-cavity instrument, particularly if the microwave source is electronically tunable over a large range, as is the case with backward wave oscillators. A number of cavity spectrometers have been designed but spectacular advances in the study of free radicals are still awaited. This should be a good method, and perhaps only requires more perseverance than it has received so far.

A particularly ingenious cavity spectrometer has been described by Radford (1968); we single it out for description since it has been used to study free radical species, although so far only species already detected by related techniques. The essential details of the frequency sweep system are illustrated in Fig. II.26. The cavity resonates in the cylindrical TE_{102} or TE_{011} modes over a frequency range from 3·5 to 7 GHz, tuning being accomplished by making one of the end walls movable so that the cylinder length is varied. The output of the microwave source is modulated at 30 MHz and the cavity is tuned to one of the 30 MHz sidebands; it is a transmission cavity and passes only the sideband to which it is tuned. This sideband is mixed with a sample of the original microwave power and the resulting 30 MHz I.F. signal is amplified and fed to a phase sensitive detector where it is compared with a reference signal from the 30 MHz oscillator. The d.c. output of the phase detector is then passed to a servo amplifier which mechanically adjusts the resonant frequency of the cavity. Frequency sweeping is thus accomplished by sweeping the source, and relying on the servo system to keep the cavity in tune. This is, of course, opposite to the AFC system described in section II.2.3 where the cavity is tuned and the source is locked to it. Radford has succeeded in detecting resonance lines from OH and OD, and his spectrometer may well prove to be successful in future attempts to detect new radical species.

II.2.6. Laser electron paramagnetic resonance

We conclude this section on electron resonance by describing some very exciting recent developments which might, in due course, render some of the

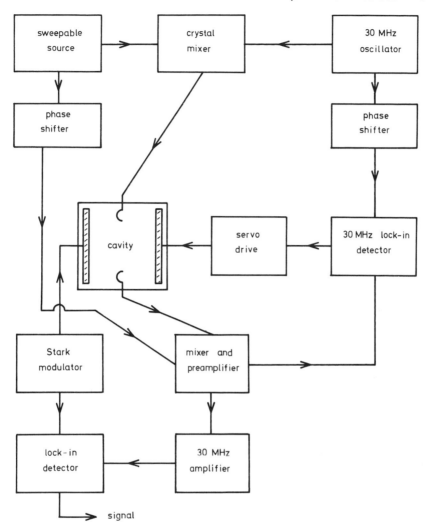

Fig. II.26. Block diagram of the tunable-cavity spectrometer built by Radford (1968).

earlier techniques obsolete. We have already pointed out that the sensitivity increases as the resonant frequency increases; precise dependence on the frequency is difficult to establish because many factors are involved, but certainly the sensitivity is at least proportional to the first power of the frequency, and perhaps more nearly proportional to the frequency squared.

Evenson and his colleagues (1968, 1971) have jumped up the frequency scale by using far-infrared lasers as radiation sources. An increasing number of laser frequencies are becoming available and although they are not usually frequency-tunable, an external magnetic field can be used to tune the energy levels of a free radical into resonance, provided the laser frequency is not too far removed from a zero-field transition frequency of the radical.

Evenson and his colleagues have so far employed HCN and H_2O lasers, and Fig. II.27 shows how the HCN laser spectrometer is used. The laser cavity (a Fabry-Perot interferometer) extends from the flat mirror on the left to the curved mirror on the right, but the lasing gas (HCN) is separated from the sample gas by a polyethylene membrane. This membrane also acts as a Brewster window which establishes the polarization of the radiation, and can be rotated so as to change the plane of polarization if required. The sample gas part of the cavity is placed between the poles of an electromagnet and, in addition, modulation coils allow the gas absorption to be modulated so that phase sensitive detection can be used. The laser beam passes through a small hole in the flat mirror and its intensity is monitored by a Golay cell. The HCN laser system operates at a frequency of 891 GHz and has been used to study rotational transitions in molecular oxygen. More recently, however, a water vapour laser with an output frequency of 2528 GHz has been used to detect resonances in OH and CH. The observation of CH represents a major development in microwave spectroscopy, since this elusive species has been sought by many spectroscopists in different parts of the world. The source of CH radicals was an oxyacetylene diffusion flame, itself an exciting development because it suggests that the high resolution spectroscopy of flames is now feasible. We shall discuss the nature of the transitions studied in Chapter IV; they are, in fact, rotational transitions which are nearly coincident with the laser frequency and can be tuned to resonance with magnetic fields in the range 0–20 kilogauss.

One can express two reservations about the general applicability of this technique. First, since laser frequencies are not at present readily tunable, one requires a near coincidence between the laser and spectroscopic transition frequencies. Second, only relatively light molecules exhibit such high frequency rotational transitions for reasonable values of N. In answer to the first point, it must be noted that laser technology is in its infancy and the effort now being expended in the development of tunable lasers will certainly alter the current situation. The second reservation has perhaps more validity but is not at present a serious limitation. In any case, there is much scope for experiments in the frequency range 100–1000 GHz and Evenson's experiments suggest that Fabry-Perot resonators might be usefully employed with more conventional microwave sources in this frequency range.

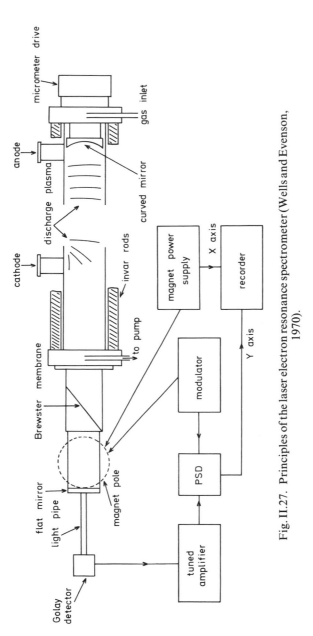

Fig. II.27. Principles of the laser electron resonance spectrometer (Wells and Evenson, 1970).

II.3. Microwave/Optical Double Resonance

The experiments described so far in this chapter have been of the direct type; they depend upon the ability to monitor the *net* absorption of microwave radiation by the molecular species of interest. Microwave/optical double resonance is an indirect method, outlined in Chapter I, in which the emission (usually fluorescence) of visible or ultraviolet radiation is monitored whilst microwave or radiofrequency transitions are induced in the molecule. Such transitions are usually concerned with the emitting excited state, and they lead to changes in the fluorescence intensity or polarization. A requirement of such experiments is that the microwave or radiofrequency transitions be induced at a rate which is sufficient to alter the populations of the emitting levels; provided this requirement is met, the sensitivity of the experiment does not depend upon whether the transitions are electric or magnetic dipole allowed, although magnetic dipole transitions have a lower probability and hence must be driven with more intense microwave or radiofrequency fields.

Excitation from the ground to the excited electronic state may be accomplished by irradiation with ultraviolet or visible light, or by electron impact. Electron irradiation has the advantages that the normal optical selection rules do not apply to the excitation, and highly energetic electronic states can be populated. It is not necessary for the detected fluorescence spectrum to involve the ground electronic state. Furthermore the microwave or radiofrequency transitions may be either magnetic (or electric) resonance transitions, or swept-frequency transitions in which an external field is not necessary. Clearly the double resonance method is potentially of wide application and many aspects have yet to be explored.

An extremely interesting precursor to more recent microwave/optical double resonance experiments is work on the CN radical, in which excitation to an excited electronic state occurs chemically; reaction of the discharge products of molecular nitrogen (which include nitrogen atoms and electronically excited N_2) with CH_2Cl_2 and other organic compounds leads to the production of the CN radical in its excited metastable $A\,^2\Pi$ state. The tenth vibrational level of this state lies very close to the $v = 0$ level of the next excited electronic state, the $B\,^2\Sigma^+$ state, and the fluorescence which accompanies the reaction of active nitrogen with CH_2Cl_2 is due to emission from the $B\,^2\Sigma^+$ state to the $X\,^2\Sigma^+$ ground state. Evenson, Dunn and Broida (1964) showed that if microwave transitions between the $A\,^2\Pi$ and $B\,^2\Sigma^+$ states were induced, the fluorescence intensity was changed. Figure II.28 illustrates the arrangement of the experiment. The active nitrogen and CH_2Cl_2 were mixed in a cylindrical TE_{011} cavity whose resonant frequency could be varied over the range 8750–10,375 MHz. The fluorescence was detected by a photo-multiplier tube, preceded by a filter which transmitted the light over the wavelength range 3200–4200 Å. The microwave power was provided by a klystron which was swept synchronously with the cavity. The microwave power was modulated at 1 kHz so that the modulated component of the fluorescence could be detected using a phase-sensitive detector. The transitions induced and the spectral lines observed are discussed in section IV.5. Even with the relatively simple detection system shown in Fig. II.28 very

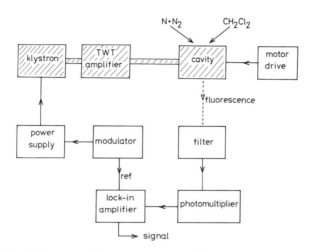

Fig. II.28. Block diagram of the apparatus used in the microwave/optical double resonance studies of the CN radical (Evenson, Dunn and Broida, 1964).

strong resonance lines were observed. There have been many subsequent refinements of the experiment and many more transitions have been detected, so that at the present time we know much more about the $A\ ^2\Pi$ and $B\ ^2\Sigma^+$ excited states of CN than we do about the ground state. Unfortunately this still appears to be the only chemical system amenable to such investigations, and later double resonance studies have employed physical methods of electronic excitation.

Silvers, Bergemann and Klemperer (1970) and Field and Bergeman (1971) have measured radiofrequency transitions in the excited $^1\Pi$ state of CS, which has a $^1\Sigma^+$ ground state. In the initial experiments the CS molecules were excited by the 2576·1 Å emission from a hollow cathode manganese lamp. This atomic line is coincident with the CS optical excitation,

$$X\ ^1\Sigma^+\ J = 7(-) \longrightarrow A\ ^1\Pi\ J = 8(+)$$

and although the upper state exhibits Λ-doublet splitting, only one of the two components is populated in the optical excitation. (The $(-)$ and $(+)$ symbols denote the parity of the state as discussed in section III.5.2.) Since the fluorescence obeys normal electric dipole selection rules, a radiofrequency field which pumps the Λ-doublet transition causes a change in the fluorescence intensity which can be monitored with the photomultiplier tube and an electrometer. In subsequent experiments Field and Bergeman showed that if the exciting light was provided by a CS molecular discharge lamp, several transitions to different rotational levels of the $^1\Pi$ state could be pumped and hence the radiofrequency spectroscopy was no longer restricted to just the $J = 8$ level. In fact the levels $J = 1$ to 9 were studied, the radiofrequency ranging from 13·6 to 1787·19 MHz. The resonance cell is illustrated in Fig. II.29; the fluorescence was monitored in a direction perpendicular to the exciting light, and the radiofrequency voltage was applied to two parallel plates, which also served as Stark electrodes. It was found to be most convenient to keep the radiofrequency constant and to sweep the energy levels through resonance by varying the static (Stark) electric field applied to the plates (see section IV.3). Figure II.30 shows a block diagram of the apparatus used by Field and Bergeman. Square wave modulation of the r.f. field intensity enabled phase-sensitive detection to be used, but even so it was usually necessary to employ extensive signal averaging in order to observe the spectrum. The Stark voltage was monitored accurately so that it was possible to determine the electric dipole moment in each rotational level. The observed line width was 2·5 MHz, which is about three times larger than would be expected from the known radiative lifetime of the $^1\Pi$ state; it seemed likely that pressure or power broadening was the reason for the observed width.

Similar experiments have been performed on the OH radical by German, Bergeman, Weinstock and Zare (1973). The ground state of OH is $^2\Pi$ but the lowest excited electronic state is the $A\ ^2\Sigma^+$ state and selective excitation is accomplished in OH through the 3072·06 Å line of discharged zinc vapour, and in OD by the 3071·60 Å line from a barium discharge glow. The excited levels studied were

$$\text{OH}\quad A\,^2\Sigma^+\quad v = 0, N = 2, J = 3/2$$

$$\text{OD}\quad A\,^2\Sigma^+\quad v = 0, N = 1, J = 3/2$$

which are further split by hyperfine interaction with the proton or deuteron. A magnetic field removes the spatial degeneracy of the hyperfine levels and the incident light is polarized perpendicular to the magnetic field direction. The fluorescence is polarized parallel ($\Delta M = 0$) or perpendicular ($\Delta M = \pm 1$) to the applied magnetic field, and a radiofrequency field applied perpendicular

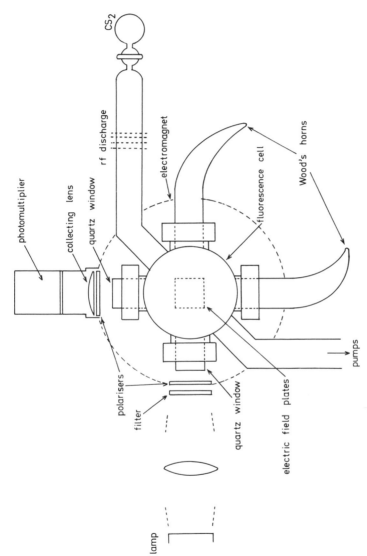

Fig. II.29. Fluorescence cell used in the microwave/optical double resonance study of CS ($A^1\Pi$) (Silvers, Bergeman and Klemperer, 1970).

to the static field induced $\Delta M = \pm 1$ transitions in the upper state, thus changing the relative intensities of the parallel and perpendicular polarized fluorescence. The frequency of the r.f. field used depends upon the magnitude of the static magnetic field, but most measurements were carried out with frequencies up to 20 MHz. The experimental arrangement was, in general, similar to that shown in Figs II.24 and II.30 except that the r.f. field was produced with a coil, rather than between parallel plates.

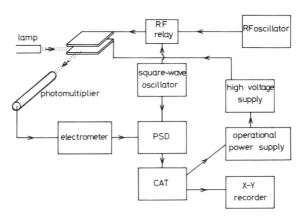

Fig. II.30. Block diagram of the apparatus used in the microwave/optical double resonance study of CS $(A^1\Pi)$.

Freund and Miller (1972) and Jost, Marechal and Lombardi (1972) have carried out similar studies of several excited electronic states of H_2, the main difference being that excitation from the ground to excited electronic state was accomplished with a beam of electrons applied parallel to a static magnetic field, rather than with ultraviolet irradiation. Figure II.31 illustrates the apparatus used by Freund and Miller. The sample gas at a pressure in the range 10^{-2}–10^{-4} torr flows into a microwave cavity situated in a strong magnetic field. The gas is subjected to electron bombardment and the fluorescence, of parallel or perpendicular polarization, is observed in directions perpendicular to the applied field. The magnetic resonance spectrum of the excited state is observed by pumping the $\Delta M = \pm 1$ microwave transitions and measuring changes in the polarized fluorescence intensity. Since a microwave cavity is used, the microwave frequency is fixed and the resonance spectrum is recorded by varying the magnetic field strength. Figure II.32 shows a block diagram of the complete system used by Freund and Miller; the microwave power is amplitude modulated at 100 kHz (by a PIN modulator)

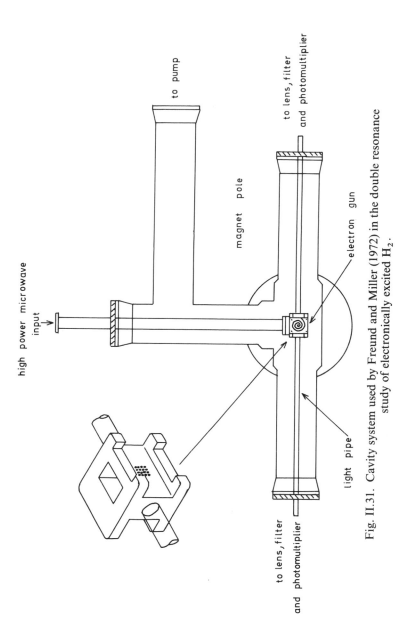

Fig. II.31. Cavity system used by Freund and Miller (1972) in the double resonance study of electronically excited H_2.

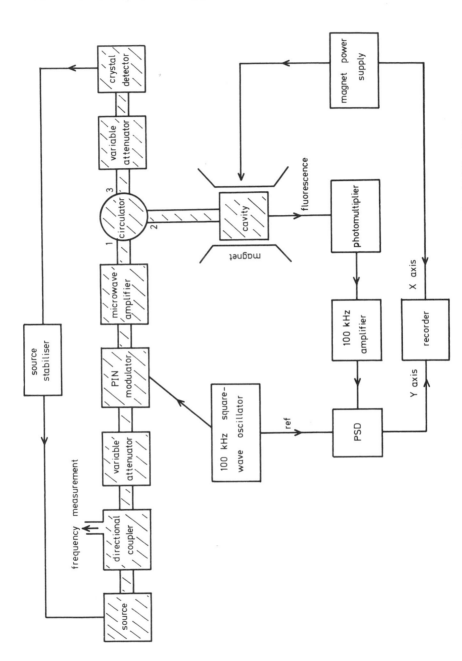

Fig. II.32. Block diagram of the complete apparatus used by Freund and Miller (1972) in the double resonance study of electronically excited H_2.

so that phase-sensitive detection can be employed. Alternatively magnetic field modulation can be used.

Details of the spectra of CS ($^1\Pi$), OH ($^2\Sigma^+$) and excited H_2 are described in Chapter IV; we also describe there a remarkable experiment by Dehmelt and Jefferts (1962) on the H_2^+ molecular ion. This is the only molecular ion to have been investigated by high-resolution spectroscopy and the experiment is again essentially a double resonance one, the main difference being that optical excitation is designed to lead to photodissociation, rather than to fluorescence. Furthermore the radiofrequency spectrum detected is that of the ground electronic state.

There seems to be little doubt that microwave or radiofrequency/optical double resonance experiments will become more numerous, and will lead to a considerable increase in our knowledge of the excited electronic states of molecules. It is, however, doubtful if such experiments will become routine, for they involve sophisticated experimental techniques, mastery of much fundamental theory, and considerable advance knowledge of the excited state, usually provided by electronic spectroscopy. It will also be interesting to see if ground states which are difficult to study by more conventional means (like that of the CH radical) will yield to double resonance experiments now in progress.

II.4. Molecular Beam Spectroscopy

II.4.1. Properties of molecular beams

A molecular beam is, as the name suggests, a well defined and fairly narrow beam of molecules moving through space in the same direction. A molecule entering an otherwise perfect vacuum travels in a straight line with constant velocity until it reaches the walls of the container. The kinetic theory of gases predicts that the mean free path of the molecules in air is about 0·3 mm at 1 torr pressure, and about 300 metres at 10^{-6} torr. The first requirement for producing a molecular beam is therefore that the background pressure be low enough for the mean free paths of the molecules to be greater than the length of the beam apparatus (typically 1 or 2 metres), and the second requirement is that the molecules forming the beam be travelling in parallel directions. A molecular beam is formed by allowing gas molecules to pass through a restricted entrance into a high vacuum chamber. Although the nature of the entrance is determined by the desire to produce as much uniform motion in a forward direction as possible, in practice the molecules emerge in directions covering a large solid angle, and moreover the turbulence outside the entrance gives rise to considerable scattering, thus further reducing the uniform directionality. Consequently it is usual to place a second selector slit or hole at such a distance from the source that only those molecules emerging within a small solid angle (and hence travelling in nearly parallel directions) are allowed to proceed further.

Molecular beams may be classified under two main headings, effusion beams or nozzle beams. An effusion beam is produced when the mean free paths of the molecules on the *high* pressure side of the entrance slit are greater than the slit dimensions. Hence ideally molecular collisions do not occur inside the entrance slit. The molecules in an effusion beam exhibit a spread in velocities which is different from the Maxwell distribution inside the source, but not too different. The average molecular velocity in an effusion beam is equal to $\frac{3}{4}(2kT\pi/M)^{\frac{1}{2}}$ where M is the mass of the molecule, so that for a beam of water at 300 K, for example, the average molecular velocity is 400 m s^{-1}. Many different kinds of entrance slit are now in use. Some workers use single rectangular slits or circular holes, others use multihole sources formed from stacks of parallel capillaries, crinkled metal foil, klystron grid stock, etc. The theory of such sources has been developed in considerable detail and the choice is determined by such factors as the required beam definition, molecular flux and, it must be admitted, personal fancy! The pressure within the source is usually a few torr, a typical single rectangular slit has an area of 10^{-2}–10^{-3} cm^2, and a typical beam flux at the detection end of the vacuum system is 10^{10}–10^{11} molecules s^{-1}. It must be emphasized, however, that these figures are only intended as a general guide since different experiments often have quite different requirements.

A nozzle beam differs from an effusion beam in a number of important respects. A requirement for its formation is that the mean free paths of the molecules inside the source be much less than the slit dimensions; consequently molecular collisions do occur inside the slit region and the resulting beam is characterized by a much more uniform forward velocity. The average molecular velocity in a nozzle beam is typically one and a half times that in an effusion beam and in a typical nozzle beam 50% of the molecules have a velocity which is within 5% of the average, whereas in an effusion beam only about 10% have velocities within 5% of the average. The adiabatic expansion in the nozzle is accompanied by transfer of rotational and vibrational energy into translational energy, resulting in important rotational and vibrational "cooling". Figure II.33 illustrates a nozzle source; the source pressure is often 100 torr or more and the beam flux is usually much higher than in an effusion beam. Since beam intensity is usually the limiting factor in molecular beam studies of all kinds, nozzle beams are finding increasing use.

The requirements for the vacuum system depend upon the nature of the beam source and the type of experiment to be performed; Fig. II.34 illustrates

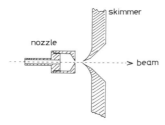

Fig. II.33. Nozzle beam source (Anderson, Andres and Fenn, 1966).

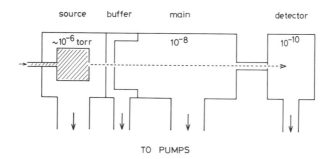

Fig. II.34. Typical four-chamber vacuum system used in molecular beam electric or magnetic resonance spectroscopy.

a typical electric resonance system, consisting of four differentially pumped chambers. The beam is formed initially in the source chamber and the background pressure during operation should be maintained at about 10^{-5}–10^{-6} torr. Rapid pumping of the source chamber is essential because this is where molecular scattering from background molecules is greatest. The beam passes through a collimating slit into a separating (buffer) chamber, which is maintained at 10^{-7} torr and serves to further define the molecular beam. The main chamber is usually the largest and is where the spectroscopic examination of the beam occurs; it will be maintained at 10^{-8} torr. Finally the beam passes through a detector slit into the detector chamber where the beam flux is monitored. Since the intention is to detect only those molecules forming the beam, the detector chamber should be at the lowest possible pressure, 10^{-10} torr or less. If the molecular species studied is readily condensable, cryogenic pumping provided by cooled surfaces is used and is very efficient. However the vacuum system is usually designed to handle gases which are not condensable at liquid nitrogen temperature.

The vacuum requirements for beam maser spectroscopy are much less exacting; a detector chamber is not required, and a separating chamber is also an unnecessary luxury. A higher molecular beam flux ($\sim 10^{14}$ molecules $\sec^{-1}$) is desirable but higher scattering due to background molecules can be tolerated; consequently typical background pressures are 10^{-5} torr in the source chamber and 10^{-6} torr in the main chamber.

As we shall see in the next section, electric and magnetic resonance experiments depend upon the ability to measure the beam intensity at the end of its passage through various static and oscillating fields, and to measure small changes in beam intensity caused by spectroscopic transitions. Many different kinds of detector have been discussed and used, but three types merit special attention here. The first of these is the surface ionization (or hot wire) detector, oxidized tungsten wire or ribbon at a high temperature (1000–2000 K) often being employed. The work function of oxidized tungsten is about 6 volts, which means that molecules with ionization potentials up to about 5·5 volts are ionized on coming into contact with the detector. The resulting ions can then be counted by the usual means, for example, with an electron multiplier tube and an electrometer. Mass analysis can also be used to ensure that only the desired ions are monitored. The surface ionization detector has been employed extensively in the detection of alkali metal atoms and compounds containing alkali metals; its efficiency can approach 100% in favourable cases.

The second and related detector is the Auger detector, which emits electrons as a result of the impact of suitably excited molecular species. Thoriated tungsten is often used, and the Auger detector is particularly good for the detection of electronically excited species, which have sufficient excess energy to stimulate electron emission from the detector surface. Again an electron multiplier tube and electrometer usually complete the detection system.

The third important detector is the so-called "universal" detector, which is in fact a mass spectrometer. The molecular beam is ionized by electron bombardment, and the resulting ion current is measured in the usual way after mass analysis to reduce the background. This method of detection can, of course, be used for any molecular beam but is restricted in its sensitivity by the relatively low ionization efficiency achieved in even the best designed systems. It is difficult to exceed 1% efficiency in the ionizer, and consequently efforts have been directed at reducing the background as much as possible. This is the reason for the high vacuum requirements of the detector chamber in Fig. II.34. Few would dispute that advances in the application of molecular beam techniques depend more on the detection efficiency than on any other single factor; the recent developments of nozzle beams and improved mass spectrometric detectors have greatly increased the scope and power of molecular beam experiments.

Having thus dealt somewhat cursorily with the formation of the molecular beam, its preservation through the suppression of background scattering, and its detection, we conclude by discussing the tricks which can be performed on the beam and which provide information about the molecular and electronic structure of the species in the beam. It is usual to subject the beam to various kinds of electric or magnetic fields, static (homogeneous or inhomogeneous) or oscillatory. The purpose of a static inhomogeneous field is to deflect or split the beam according to the possible spatial orientations of the electric or magnetic moment of the molecule, whilst the purpose of the oscillating field is to induce spectroscopic transitions which change the spatial orientation of the molecule and hence its response to a static field. The explicit purposes will become clearer in the following sections when we deal with specific examples, but there are a number of general features which can be discussed here. It is worth noting that in nearly all experiments the fields used, static or oscillating, are all electric or all magnetic, although there are a few situations where both electric and magnetic fields can be used. To deal with magnetic fields first, a molecule in a beam passing through a magnetic field gradient whose direction is perpendicular to the beam experiences a force which induces a transverse acceleration (a) given by

$$a = (\mu_{\text{eff}}/M)(\partial B/\partial z) \tag{II.9}$$

where M is the mass of the molecule, μ_{eff} is its effective magnetic moment, and $(\partial B/\partial z)$ is the field gradient. Figure II.35(a) illustrates a type of dipole magnet

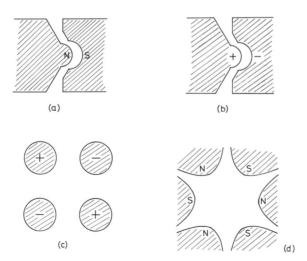

Fig. II.35. Pole configurations for: (a) magnetic dipole field; (b) electric dipole field; (c) electric quadrupole field; (d) magnetic hexapole field.

system often used to produce the field gradient. Now the important point is that for any particular molecule, the magnitude and sign of μ_{eff} depend upon the quantum state, or spatial orientation, of the molecule. For example, in the case of an atom in a ^2S state, the magnetic moment arises from electron spin and there are two possible orientations of the spin in the magnetic field. Consequently there are two possible values of μ_{eff}, given by

$$\mu_{eff} = \pm \tfrac{1}{2} g_e \mu_B \tag{II.10}$$

where, as we shall see, g_e has a value very close to 2, and μ_B is the Bohr magneton. Hence some atoms experience a "positive" transverse acceleration, others experience a "negative" transverse acceleration, and consequently the beam splits into two. This indeed, was exactly the classic experiment first performed on a beam of silver atoms by Stern and Gerlach (1924) which demonstrated the property of an intrinsic magnetic moment arising from electron spin. Now we might choose to arrange the apparatus so that only one component of the split beam is allowed to proceed, in which case we can conduct experiments on a "polarized" beam, that is, one in which ideally all the atoms have the same spin (spatial) orientation. Similar experiments can be performed with a dipole electric field, the geometry of which is illustrated in Fig. II.35(b). In this case the transverse acceleration of a molecule passing through the electric field gradient is given by

$$a = (\mu_{eff}(M)(\partial E/\partial z) \tag{II.11}$$

where μ_{eff} is now the effective electric moment of the molecule. It is, of course, necessary that the molecule possess an electric dipole moment in order for this interaction to occur. It must also be pointed out that the number of allowed orientations of the magnetic or electric moment is not necessarily two; the number of possible spatial orientations actually depends upon the interplay of the electron spin, nuclear spin, rotational angular momentum, electronic orbital angular momentum, etc. within the molecule. We shall discuss the details later, particularly in Chapters III and IV; suffice it to say here that the function of the inhomogeneous field is to isolate desired quantum states of the molecules forming the beam through the effects on their trajectories in the apparatus. In other words, non-thermal population distributions of the types discussed in Chapter I are produced.

We are not restricted to dipole fields and in many experiments it is advantageous to use more complex field configurations in order to achieve the desired state selection. Quadrupole electric fields produced by charged rods in the configuration illustrated in Fig. II.35(c) are particularly useful in those cases where the Stark effect is second-order, and they have the advantage of producing both state separation and some space focussing of the beam. The quadrupole has zero electric field along the central axis of the four rods and a large field in the vicinity of the rods. Consequently state separation arises from the fact that those molecules whose energy increases in an electric field tend to be concentrated towards the axis, where the field is small, whilst molecules whose energy decreases in an electric field are attracted towards the rods and hence removed from the beam. Similar state separation and focussing is produced by a hexapole magnetic field (Fig. II.35(d)) for molecules or atoms with a magnetic moment and first-order (linear) Zeeman effect.

Finally we come to the region of the apparatus in which the spectroscopic transitions are induced. In many cases the spectrum studied is a resonance spectrum in which the separation of the energy levels is determined by a *homogeneous* magnetic or electric field, and transitions between them are induced by an appropriate radiofrequency or microwave field. In other cases, notably beam maser spectroscopy, the transitions induced are of exactly the same type as those in pure microwave spectroscopy; the frequency is swept and an applied static field is not required.

The great advantage of molecular beam spectroscopy is the very high resolution which is obtainable. The absence of collisions between molecules means that pressure broadening is absent, and since the electromagnetic radiation is usually applied in a direction perpendicular to the beam, Doppler broadening is also removed. Assuming that instrumental contributions are negligible, the line width is determined by the time the molecules spend in the radiation field. This time depends on the molecular velocity and on the length of the radiation field, both of which can be controlled to some extent, particularly the length of the radiation field. In experiments with homogeneous static fields, however, the line width is often determined by field inhomogeneities.

II.4.2. Molecular beam magnetic and electric resonance spectroscopy

The first experiment on an excited electronic state using molecular beam magnetic resonance was described by Lichten (1960, 1962, 1971) who made a thorough study of the H_2 molecule in its excited c $^3\Pi_u$ state. The details of the transitions studied and their analysis are given in section IV.8, and here we confine attention to the purely experimental aspects. Similar experiments have been carried out on the A $^3\Sigma_u^+$ excited state of N_2. Figure II.36 applies equally to both studies and illustrates the trajectory of a typical molecule as it passes through a series of magnetic fields. In Lichten's experiments the H_2 molecules are first excited into the c $^3\Pi_u$ state by electron bombardment, and the beam then passes through a dipole magnetic field (A), a homogeneous magnetic field (C) and a second dipole field (B). The A dipole field deflects a molecule whose trajectory and effective magnetic moment are such that it would otherwise leave the beam. This molecule is then deflected by the B field, which has opposite polarity, such that it hits a stop wire and is scattered. If, however, a resonance transition is induced in the homogeneous C field by means of an appropriate radiofrequency or microwave field, the effective magnetic moment of the molecule is changed, it no longer satisfies the collimating condition imposed by the B field, and hence misses the stop wire and reaches the detector. Consequently resonance transitions result in an increase in the number of molecules reaching the detector. In the first experiments on excited H_2 the detector used was a nickel secondary electron emission detector combined with an electrometer, but in later experiments (1971) a thoriated tungsten detector was used. The radiofrequency transitions were induced by passing the beam through an r.f. coil, the transition region being typically 6 mm in length.

Similar experiments have been carried out recently by Freund, Miller, De Santis and Lurio (1970) on the metastable excited $A\ ^3\Sigma_u^+$ state of N_2. A molecular beam of N_2 is irradiated with electrons to produce the $A\ ^3\Sigma_u^+$ molecules, which are again monitored with an Auger detector. The A and B dipole fields are in the range 1–2 kilogauss, and the ratio of the field gradient to field is about 3 cm^{-1}. Resonance transitions with frequencies up to about 32 MHz are detected, with homogeneous C fields up to 20 gauss being used. Again the details of the transitions studied are given in section IV.7.

Molecular beam electric resonance studies on LiO (a ground state free radical) and CO in its excited metastable a $^3\Pi$ state have been carried out by Klemperer and his colleagues (1965, 1970, 1971, 1972). Figure II.37 shows a

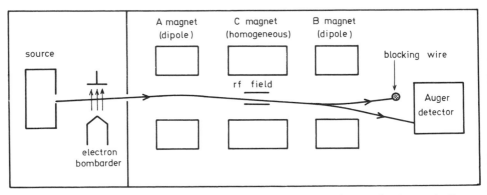

Fig. II.36. Molecular beam magnetic resonance system used to study electronically excited H_2 and N_2.

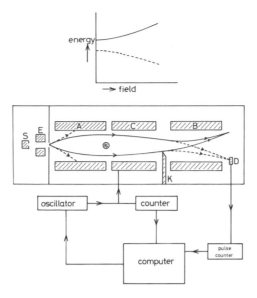

Fig. II.37. Molecular beam electric resonance spectrometer.

block diagram of a typical electric resonance spectrometer, the main difference between this and that shown in Fig. II.36 being that the A, B and C fields are now all *electric* fields. In most cases the A field is a quadrupole field, but the analysing B field may be either a quadrupole field, or alternatively a dipole field. The molecular trajectories shown in Fig. II.37 refer to molecules in one or other of two energy levels which can be mixed by an applied electric field to show a second-order Stark effect. As we discussed in the previous section, molecules in the upper state (continuous line) are focussed by the quadrupole A field, whilst those in the lower state (dotted line) are removed from the beam. The B field can then be arranged to deflect upper state molecules *away* from the detector. If, however, radiofrequency or microwave transitions are induced in the homogeneous C field region, the number of lower state molecules increases, and these can be deflected onto the detector by the B field. Spectroscopic transitions thus result in an increased detector current.

Freund, Herbst, Mariella and Klemperer (1972) have studied LiO, the beam source being an iridium tube oven in which Li_2O is heated to 1800 K. Measurements were made on the isotopic species ^7LiO which, although it formed only a small percentage of the total beam, was nevertheless detectable with a surface ionization detector. Freund *et al.* used a dipole B field, and operated in the "flop-in" mode, that is, resonance transitions resulted in an increased beam intensity at the director and hence an increased electron current. In earlier experiments Freund and Klemperer (1965) studied CO in its excited a $^3\Pi$ state, excitation being accomplished by electron bombardment of the beam. Again they used a quadrupole A field, a dipole B field, and an Auger detector consisting of a freshly deposited sodium surface. Many different electric resonance transitions were studied, and the results are described in detail in section IV.8.

Although the number of molecular beam resonance studies on open shell molecules is, as yet, rather small, the methods are clearly of wide application and are particularly powerful for the study of molecules in metastable excited electronic states.

II.4.3. Molecular beam maser spectroscopy

The experiments described in the previous section are *indirect*, in that the absorption of radiofrequency or microwave energy is detected through its effect on the molecular trajectory. In molecular beam maser spectroscopy, however, we return to the direct detection of the emission or absorption of radiation. Conventional absorption experiments of the type discussed in section II.1 and section II.2 are not, in general, sensitive enough to be combined with molecular beam techniques, simply because the number of molecules in a beam is rather small. We recall, however, that in normal absorption spectroscopy we depend on the ability to measure a *net* absorption of radiation, whose magnitude depends upon the relative populations of the two energy levels involved in the transition. For radiofrequency or microwave transitions the population difference for a gas at thermal equilibrium is very small at normal temperatures. Beam maser spectroscopy depends upon the ability to create artificially a *large* population difference, usually a population inversion. As we have seen in section II.4.1, the quadrupole electrostatic field achieves just this for levels which are coupled through a Stark effect. In a maser spectrometer, therefore, the C and B field regions are usually replaced by a microwave cavity, which is tuned to the transition frequency. Microwave power of the correct frequency is fed to the cavity, stimulating the transitions which result in the emission of more radiation. In other words, the input microwave power is amplified, and the spectrometer system is arranged so that the output signal is detected. As mentioned earlier, the molecular beam does not need to be so well defined as in the resonance deflection experiments, so that wider beams of greater molecular flux can be used. Figure II.38 illustrates the principal features of beam maser spectroscopy using a microwave cavity. Masers have, however, been operated at frequencies as low as a few MHz, the cavity being replaced by a parallel plate condenser. It is also possible to focus lower state

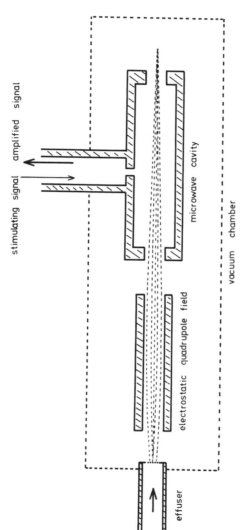

Fig. II.38. Principles of a molecular beam maser spectrometer.

molecules and hence to produce a population cooling. In this case, of course, stimulated absorption is observed.

The microwave emission signal is usually detected by superheterodyne techniques, a typical intermediate frequency being 30 MHz. Figure II.39 shows a block diagram of the microwave and electronic units in a typical beam maser spectrometer. The microwave power fed to the cavity at frequency f is necessarily at a very low level (typically 10^{-10}–10^{-12} watts). The local oscillator source at frequency $(f + 30)$ MHz may be a separate oscillator, or may be derived from the primary source by generation of 30 MHz sidebands. The 30 MHz signal from the mixer is amplified and detected in the normal way. It is also common practice to chop the beam mechanically with a rotating fan blade, or to frequency modulate the stimulating source, so that the output microwave signal is amplitude modulated. Consequently the output from the 30 MHz detector is fed to a phase-sensitive detector where it is compared with a modulation frequency reference signal. The d.c. output signal from the phase-sensitive detector can then be displayed on an oscilloscope or an X–Y recorder.

The microwave frequency is swept, but necessarily only over a very small range because, as we saw earlier in this chapter, the bandwidth of a high-Q cavity is of the order of 1 MHz. Typical sweep amplitudes in beam maser spectroscopy are therefore a few hundred kilohertz at most, the line widths usually being in the range 1–10 kHz. Consequently beam maser spectroscopy

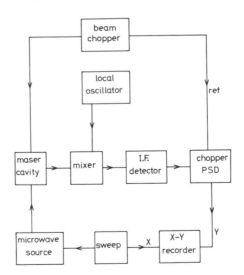

Fig. II.39. Block diagram of the microwave and electronic units in a typical molecular beam maser spectrometer.

is only a practical proposition if the transition frequency is already known fairly accurately, usually from conventional microwave measurements. The main purpose of beam maser experiments, therefore, is to improve the precision of the measurements, typically by two orders of magnitude, and to resolve splittings which are too small to appear in conventional microwave spectroscopy.

Only one free radical has, as yet, been studied by beam maser spectroscopy. This is the OH radical, detected and measured by Ter Meulen and Dymanus (1972). The OH radical was produced by reaction of H atoms with NO_2 inside the source, and the molecular beam contained a sufficiently high OH concentration for its maser spectrum to be detected. As we shall see in Chapter IV, the OH radical has an extensive microwave spectrum and Ter Meulen and Dymanus remeasured a number of transition frequencies, the details of which are described in Chapter IV. This work is important, not only for its intrinsic interest, but also for the hope it encourages that other free radicals may be studied by beam maser techniques.

II.5. Radioastronomy

Some of the most interesting microwave spectra of free radicals have been provided by radioastronomers during the last few years, and it seems likely that more species which are elusive in the laboratory may be detected in interstellar gas clouds. There are, indeed, already a number of unidentified microwave lines from interstellar clouds and we therefore conclude this review of experimental methods with a very brief and elementary description of the techniques used by radioastronomers to record the microwave spectra of interstellar molecules.

The interstellar microwave spectra may be detected either as emission or absorption spectra. Most interstellar molecules have been detected through the spontaneous emission of radiation which accompanies a transition from an upper to a lower state. Despite the fact that at microwave frequencies the probability of such transitions is low, the radiotelescopes are sensitive enough for these spectra to be detected. In a few cases, stimulated emission is observed (notably from OH and H_2O), the initial stimulating source being the background 2·7 K black-body radiation which is everywhere present. Absorption spectra have also been recorded, either against the background radiation, or more often against continuous radiation emitted from another more distant source.

Many different types of radiotelescope are now in use, but Fig. II.40 illustrates a typical system. The radiation is collected by a large parabolic dish and focussed at the detector, to which we return later. The dish is ideally

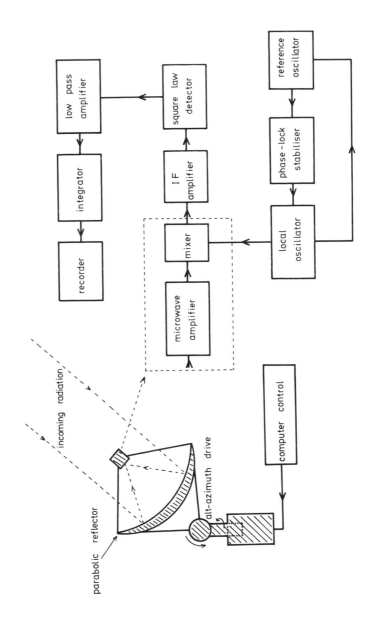

Fig. II.40. Block diagram of a typical radiotelescope system.

as large as is consistent with other requirements. Steerable dishes up to 350 feet in diameter have been built, the virtues of size being the collection of the maximum amount of radiation and the optimization of the angular resolution of the interstellar source. However size produces its problems, quite apart from that of its cost. If an incoming signal is not to be degraded, the surface of the dish must be uniform in comparison with the wavelength of the radiation collected. Bearing in mind that the detected radiation is often of wavelengths down to a few millimetres, this requirement poses considerable engineering problems, for radiotelescope dishes are necessarily exposed to variations in the terrestial elements (e.g. the weather) and subject to deformations arising from their massive size and weight. Fortunately the shorter the wavelength of the detected radiation, the smaller is the required dish diameter to achieve the same spatial resolution. Many of the interstellar molecules have been detected with a telescope at Kitt Peak in Arizona which has a dish diameter of only 36 feet. Another important requirement of the dish is that its orientation be adjustable. Figure II.40 illustrates the fact that two degrees of freedom, where one is rotation about an axis and the other is a rotation in a plane containing the axis, are sufficient to allow the telescope to be pointed in any direction.

The detector which is placed at the focus of the dish will usually consist of a microwave amplifier to boost the incoming signal, and a mixer in which the interstellar radiation is mixed with a standard laboratory generated local oscillator frequency to produce an intermediate beat frequency. In some telescopes intermediate frequencies (I.F.) above 1 GHz are now used, there being less interference at these higher frequencies. Once the incoming interstellar signal has been reduced to an I.F. signal, the subsequent processing takes place in conventional ground-based laboratories; the I.F. signal is amplified, detected with a square-law detector and displayed. Various methods can be used to modulate the detected signal, including mechanical chopping of the microwaves in the detector, and even vibration of the telescope dish! Of course, the frequency of the incoming radiation cannot be swept and for a given local oscillator frequency the range scanned by the telescope is determined by the bandwidth of the detector system. Typical maximum bandwidths are a few tens of megahertz at most, and for this reason a knowledge of the microwave transition frequencies is almost essential for the successful detection and identification of any particular molecule. It should also be noted that the interstellar frequencies are shifted somewhat from the laboratory rest frequencies because of the Doppler effect.

Many other kinds of detector system have been used, and the single receiver dish is often replaced by two or more smaller dishes. Kraus (1966) has written an admirable book which reviews the experimental techniques used in radioastronomy and the laboratory spectroscopist can learn much by familiarizing himself or herself with these techniques. The work on interstellar molecules, carried out mainly by scientists in the United States, is the product of a stimulating fusion of the interests between the astronomers and spectroscopists.

III. Theory of Molecular Energy Levels

III.1. Introduction

Spectroscopists interpret their data at many different levels of sophistication. At one extreme we find the purely empirical approach; the appearance of certain groups of infrared absorption bands, for example, may be associated with particular groupings of atoms in molecules. Such an interpretation does not involve discussion of energy levels, spectroscopic transitions, or even molecular constants. Many of the analytical applications of spectroscopy in chemistry are discussed at this level.

The next level of interpretation is probably in terms of expressions for energy levels, coupled with rules for deciding which transitions between the levels are observable. For example, the microwave rotational spectrum of CO can be interpreted by saying that the lines arise from transitions between rotational levels of the molecule whose energies are given by the equation

$$E_{rot} = BJ(J + 1). \qquad \text{(III.1)}$$

B is called the rotational constant and J is called the rotational quantum number, taking the values 0, 1, 2, 3, etc. We would also say that only transitions between adjacent levels can occur, and that therefore the "allowed" transitions obey the "selection rule" $\Delta J = \pm 1$. The first few lowest rotational levels and their energies are

$$
\begin{aligned}
J &= 0, & E_{rot} &= 0 \\
J &= 1, & E_{rot} &= 2B \\
J &= 2, & E_{rot} &= 6B \\
J &= 3, & E_{rot} &= 12B \quad \text{etc.}
\end{aligned}
\qquad \text{(III.2)}
$$

and the rotational transitions and frequencies are

$$J = 0 \rightarrow J = 1, \qquad v = 2B$$
$$J = 1 \rightarrow J = 2, \qquad v = 4B \qquad\qquad \text{(III.3)}$$
$$J = 2 \rightarrow J = 3, \qquad v = 6B \quad \text{etc.}$$

Hence measurements of the frequencies of the lines give a value for the rotational constant B.

Now some might well be satisfied at this stage since the spectrum is, up to a point, "interpreted". There are, nevertheless, a number of important questions which should be asked:

1. Why are the rotational levels given by (III.1)?
2. What is the rotational constant B?
3. What is the quantum number J?
4. What are rotational levels anyway?
5. Why is the selection rule $\Delta J = \pm 1$ obeyed? Are molecules always so obedient?
6. Can we be certain that (III.1) is correct, even if it seems to fit the experimental results?

These questions are related and once we attempt to answer them properly, we are in at the deep end. There is an answer to question 2 sufficient for some purposes, which is to assert that the molecule can be regarded as consisting of two nuclei held rigidly in position by the electrons. The rotational constant B is then inversely proportional to the moment of inertia (I) of the molecule and is, in fact, given by

$$B = \frac{h}{8\pi^2 I}. \qquad\qquad \text{(III.4)}$$

I depends on the masses m_1 and m_2 of the nuclei and r, the distance between them, according to the equation

$$I = \frac{m_1 m_2}{m_1 + m_2} r^2. \qquad\qquad \text{(III.5)}$$

If we determine B from the spectrum and look up the values of the masses m_1 and m_2, we can determine r, the "bond length".

Now we find that equation (III.1) does not fit the experimental results perfectly, and it is not difficult to raise objections to the model proposed. The most obvious is that molecules are not rigid, but vibrate. One can start adding correction terms to (III.1), the first refinement being the equation

$$E_{\text{rot}} = BJ(J + 1) - DJ^2(J + 1)^2. \tag{III.6}$$

Of course, this equation (which is well founded) is bound to fit the data better than (III.1) because it contains two constants, B and D, the new constant being called a centrifugal distortion constant. We can go on adding more terms to (III.6) which improve the fit with experiment still more. However one must soon reach a stage where doubts occur. Even if an expression with several constants does fit the data, are we certain that the expression is correct and that the constants are meaningful? In the CO "rotational" spectrum, can we be sure that the electrons play no role other than to prevent the nuclei from flying apart? When the nuclei rotate, for example, do the electrons keep up with them? There is a reassuring approximation in the background, called the Born-Oppenheimer approximation, which states that because of the difference between nuclear and electronic masses, we are justified in treating nuclear motion separately from electronic motion. It works very well for many experiments on many molecules. But if we are measuring very narrow lines and quoting constants like B to very high accuracy, it might need looking at again. In fact the rotational levels of the hydroxyl radical are certainly not given by (III.6), for many reasons. The radical has unpaired electron spin, electronic orbital angular momentum, proton nuclear spin, and it rotates very rapidly because the hydrogen atom is so light.

The reader will appreciate that we have not really begun to answer the questions posed about equation (III.1), and although we can always "interpret" spectra by adding seemingly appropriate terms to the expressions for the energy levels and determining the values of the constants, we are on dangerous ground. There is little point in quoting values for constants until we can say exactly what the constants are and where they come from. The difficulty is that once curiosity is aroused, it is not easily satisfied. Whatever level of understanding we reach, there is seemingly always another yet deeper level and if we dig deep enough, we soon reach the current limits of understanding, which is where the theoretical physicist is earning a living. So although it is important to realize what the limitations are, one always reaches a point where the desire to solve more immediate problems must compromise with curiosity about the foundations. The exact point of compromise is a matter for individual choice.

In the following section, therefore, we will attempt to trace the steps which lead to reliable expressions for the energy levels of interest in the experimental results to be described later. We shall attempt to do so using the bare minimum of quantum mechanics although this is an unsatisfactory compromise.

The best that can be hoped for is that this chapter will interest the reader, give him (or her) some idea of the purpose and interpretation of the experiments, and leave him dissatisfied but forgiving. The next step then is to master the technicalities of the language of quantum mechanics, but for this the reader must look elsewhere.

III.2. The Molecular Hamiltonian

III.2.1. Introduction

The logical development of the theory would be to start with the fundamental principles and finish with expressions for the spectral transition frequencies. In this section we will indeed tackle the problem this way, but since it is often much easier to appreciate the development of a theory if one knows something about the final goal, let us first briefly trace the process in reverse as follows:

(a) The microwave spectrum is recorded and yields a set of transition frequencies together with relative line intensities.

(b) The experimental data are interpreted in terms of expressions for the relevant energy levels and appropriate spectroscopic selection rules.

(c) The energy levels are derived theoretically by constructing a so-called "effective Hamiltonian" and finding its eigenvalues. The effective Hamiltonian is an expression summarizing the magnetic and electrostatic interactions which are significant in the particular example. The various terms in this expression usually consist of products of angular momentum operators multiplied by molecular constants.

(d) The "effective Hamiltonian" is derived from what we will call the "complete Hamiltonian" usually by some sort of perturbation procedure. The complete Hamiltonian contains terms which describe explicitly all the kinetic, potential, and spin energy contributions of the electrons and nuclei to the total energy of the molecule. The complete Hamiltonian also contains terms describing interactions with external fields. In contrast, the effective Hamiltonian will usually only refer *explicitly* to one or two contributions to the energy, the effects of all the other contributions being contained in the molecular constants. For example, an effective *rotational* Hamiltonian will usually contain terms referring explicitly to the rotational angular momentum of the nuclei and to the electronic and nuclear spin, but the effects of electronic and vibrational energy are usually contained in the molecular constants appearing in the effective Hamiltonian.

(e) The complete Hamiltonian is derived by starting with a general expression for the kinetic and potential energy of the electrons and nuclei and carrying out a series of coordinate transformations designed to separate, as far as possible, the translational, vibrational, rotational, and electronic contributions to the energy. Complete separation is not possible, and therefore important cross-terms between the different types of motion exist. Magnetic interactions involving electron and nuclear spins and external magnetic fields are derived separately starting from relativistic quantum mechanics, and again using appropriate coordinate transformations.

This all sounds fairly complicated, and so it is! It is not difficult to follow, however, and this logical development is the only alternative to pulling terms out of a hat. In particular, an understanding of the procedure is vital if one is to gain insight into the nature of the molecular constants, which are the end product of the experiment.

III.2.2. Hamiltonians, eigenfunctions and eigenvalues

As a prerequisite to pursuing the development of the molecular Hamiltonian, it is essential to understand something of the formalism and methods of quantum mechanics, particularly that part relating to Hamiltonian operators, their eigenfunctions and their eigenvalues. The reader who is new to this might find it puzzling, but can be assured that the language is assimilated with use, and appreciation of the subtleties follows later.

We shall find it convenient to use a notation due to Dirac, and perhaps a few words of clarification and definition might help the reader who is new to this notation. Many readers will be familiar with the concept that the state of a system may be described by a function $\psi(\mathbf{r}_1, \mathbf{r}_2, \ldots, t, \ldots)$ where $\mathbf{r}_1, \mathbf{r}_2$, etc. are the position vectors of the particles in the system, t is time, and so on. Very often we characterize the state of a system by specifying the values of various relevant "quantum numbers", p, q, r, etc. so that in this method of nomenclature we describe the state in terms of a function

$$\psi_{p,q,r,\ldots}(\mathbf{r}_1, \mathbf{r}_2, \ldots, t, \ldots).$$

In the Dirac notation the state is described by the symbol

$$|p, q, r, \ldots\rangle$$

which is called a "ket". The conjugate function $\psi^*_{p,q,r,\ldots}(\mathbf{r}_1, \mathbf{r}_2, \ldots, t, \ldots)$ is associated with the Dirac symbol

$$\langle p, q, r, \ldots|$$

called a "bra". Dirac bras and kets describe states and are somewhat akin to vectors in that they make no reference to a particular choice of coordinate system; in contrast, wave functions are defined in a particular coordinate system. The number of quantum numbers actually specified in a ket depends upon the particular problem considered.

Now in the conventional notation the normalization condition is expressed by means of the integral requirement

$$\int_{\text{all space}} \psi_p^*(\mathbf{r})\psi_p(\mathbf{r})\, d\tau = 1$$

which in Dirac notation is written

$$\langle p \mid p \rangle = 1.$$

The expectation value of an operator $\hat{\mathbf{O}}$ for the state $\psi_p(\mathbf{r})$ in the two notations is

$$\int \psi_p^*(\mathbf{r})\hat{\mathbf{O}}\psi_p(\mathbf{r})\, d\tau \equiv \langle p|\hat{\mathbf{O}}|p \rangle \tag{III.7}$$

which is an example of the more general relationship

$$\int \psi_p^*(\mathbf{r})\hat{\mathbf{O}}\psi_{p'}(\mathbf{r})\, d\tau \equiv \langle p|\hat{\mathbf{O}}|p' \rangle. \tag{III.8}$$

Other examples of the Dirac notation will follow in this chapter; it is perhaps worth noting that although the algebraic manipulation of Dirac symbols is a highly developed and sophisticated subject, we shall not need to make use of the subtleties in this book.

We denote the effective Hamiltonian referred to in section III.2.1 by the symbol $\mathcal{H}_{\text{eff}}$; the eigenfunctions $|\psi_1\rangle$, $|\psi_2\rangle$, etc. of this Hamiltonian are defined by the operator equations

$$\mathcal{H}_{\text{eff}}|\psi_1\rangle = E_1|\psi_1\rangle$$
$$\mathcal{H}_{\text{eff}}|\psi_2\rangle = E_2|\psi_2\rangle \tag{III.9}$$

$$\text{etc.}$$

The first equation, for example, means that if we operate on a function $|\psi_1\rangle$ with the Hamiltonian $\mathcal{H}_{\text{eff}}$, we obtain the same function $|\psi_1\rangle$, multiplied by E_1 which is an eigenvalue of the Hamiltonian $\mathcal{H}_{\text{eff}}$. Of course, equation (III.9) represents the successful conclusion of the theory, since the E_1, E_2, etc. are the energy levels referred to in section III.2.1(b) which we seek. In practice we start by knowing the appropriate Hamiltonian $\mathcal{H}_{\text{eff}}$, but we do not know what the final eigenfunctions $|\psi_1\rangle$, $|\psi_2\rangle$, etc. are and the task is to find them and their energies. The procedure normally followed is to choose a set of orthonormal functions, called the basis set, on which we operate with

$\mathcal{H}_{\text{eff}}$. Let us call these basis functions $|\phi_1\rangle$, $|\phi_2\rangle$, $|\phi_3\rangle$, etc. Then in general, when we operate on a particular member of the set, say $|\phi_1\rangle$, with $\mathcal{H}_{\text{eff}}$, we obtain the result

$$\mathcal{H}_{\text{eff}}|\phi_1\rangle = k_{11}|\phi_1\rangle + k_{21}|\phi_2\rangle + k_{31}|\phi_3\rangle + \cdots. \qquad \text{(III.10)}$$

As a consequence of equation (III.10) we may say that the value of the matrix element of $\mathcal{H}_{\text{eff}}$ connecting the states $|\phi_1\rangle$ and $|\phi_2\rangle$ is k_{21}; we write this statement in the form

$$\langle\phi_2|\mathcal{H}_{\text{eff}}|\phi_1\rangle = k_{21}. \qquad \text{(III.11)}$$

Now why is the left-hand side of (III.11) called a "matrix element"? The reason is that we can summarize the results of operating with $\mathcal{H}_{\text{eff}}$ on the basis functions $|\phi_1\rangle$, $|\phi_2\rangle$, etc. in the form of a matrix as follows:

$$\begin{array}{c c c c c}
 & |\phi_1\rangle & |\phi_2\rangle & |\phi_3\rangle & \cdots \\
\langle\phi_1| & k_{11} & k_{12} & k_{13} & \cdots \\
\langle\phi_2| & k_{21} & k_{22} & k_{23} & \cdots \\
\langle\phi_3| & k_{31} & k_{32} & k_{33} & \cdots \\
 & \vdots & \vdots & \vdots & \vdots & \vdots
\end{array}. \qquad \text{(III.12)}$$

Equation (III.11) thus represents the entry in the second row, first column of this matrix. The desired final results, given in (III.7), can also be written as a matrix:

$$\begin{array}{c c c c c}
 & |\psi_1\rangle & |\psi_2\rangle & |\psi_3\rangle & \cdots \\
\langle\psi_1| & E_1 & 0 & 0 & \cdots \\
\langle\psi_2| & 0 & E_2 & 0 & \cdots \\
\langle\psi_3| & 0 & 0 & E_3 & \cdots \\
 & \vdots & \vdots & \vdots & \vdots & \vdots
\end{array}. \qquad \text{(III.13)}$$

However, the obvious and important difference between these two matrices is that in (III.13) all the off-diagonal elements are zero; in other words, the matrix is diagonal. The mathematical part of the exercise is therefore to convert the matrix (III.12), in which some or all of the off-diagonal elements are non-zero, into the diagonal matrix (III.13). When we have achieved this we say that the functions $|\psi_1\rangle$, $|\psi_2\rangle$, etc. diagonalize the Hamiltonian $\mathcal{H}_{\text{eff}}$, that is, they are eigenfunctions of $\mathcal{H}_{\text{eff}}$ with eigenvalues E_1, E_2, etc. One finds that the eigenfunctions can be represented as linear combinations of the basis functions, e.g.

$$|\psi_1\rangle = \sum_n c_n |\phi_n\rangle \qquad (III.14)$$

and the eigenvalues E_1, E_2, etc. can be expressed in terms of the quantities k_{11}, k_{12}, etc. The process of diagonalizing the matrix (III.12) is made much easier if we choose the basis functions $|\phi_1\rangle$, $|\phi_2\rangle$, etc. such that the off-diagonal elements are as small as possible. The reader may well wonder what the functions $|\phi_1\rangle$, $|\phi_2\rangle$, etc. look like. As we shall see later, it is not usually necessary to use specific functional forms, but merely to specify for each state $|\phi_i\rangle$ a set of appropriate quantum numbers. Operation on the state with $\mathcal{H}_{\text{eff}}$ then results in changes in some or all of these quantum numbers, yielding a different state $|\phi_j\rangle$ or linear combination of such states. As we shall see in section III.4, an appreciation of the main physical features of the molecule usually helps one to choose the most sensible basis set.

In our later discussions we shall frequently refer to interactions as being first-order, second-order, etc. and we should perhaps offer a word or two of explanation of these terms which come from perturbation theory. Suppose that our Hamiltonian consists of a large term $\mathcal{H}^{(1)}$ and a smaller term V, and that $|\phi_1\rangle$, $|\phi_2\rangle$, etc. are eigenfunctions of $\mathcal{H}^{(1)}$ with energies ε_1, ε_2, etc. If V has non-vanishing matrix elements between the basis functions, the matrix of the Hamiltonian will take the form,

$$
\begin{array}{c|cccc}
 & |\phi_1\rangle & |\phi_2\rangle & |\phi_3\rangle & \cdots \\
\hline
\langle\phi_1| & \varepsilon_1 + V_{11} & V_{12} & V_{13} & \cdots \\
\langle\phi_2| & V_{21} & \varepsilon_2 + V_{22} & V_{23} & \cdots \\
\langle\phi_3| & V_{31} & V_{32} & \varepsilon_3 + V_{33} & \cdots \\
\vdots & \vdots & \vdots & \vdots & \vdots
\end{array}
\qquad (III.15)
$$

Now the initial energy of state $|\phi_1\rangle$ is modified by the perturbation V, and the modified energy, E_1, may be expressed as a power series expansion of the form

$$E_1 = \varepsilon_1 + V_{11} + \sum_{j \neq 1} \frac{V_{1j} V_{j1}}{\varepsilon_1 - \varepsilon_j} + \cdots \qquad (III.16)$$

whilst the wave function $|\phi_1\rangle$ is correspondingly modified to a new function $|\psi_1\rangle$ according to the expansion

$$|\psi_1\rangle = |\phi_1\rangle + \sum_{j \neq 1} \frac{V_{1j}}{\varepsilon_1 - \varepsilon_j} |\phi_j\rangle + \cdots . \qquad (III.17)$$

The terms on the right-hand side of (III.16) are called the zero-order, first-order, second-order, etc. terms respectively. Although the numerator in the second-order term consists of a matrix element of V multiplied by its conjugate element, it might be the case that V consists of a sum of two or more terms, i.e., $V = V^{(1)} + V^{(2)} + \cdots$ in which case the second-order term in (III.16) would then consist of products of matrix elements of the different terms $V^{(1)}$, $V^{(2)}$, etc. This is what we mean when we talk about cross-terms between different interactions, particularly in section III.2.7.

Having introduced some of the language of quantum mechanics here, we conclude this section by describing what we shall *not* do. We shall not discuss the details of the operator algebra, how to construct the matrix (III.12), or how to diagonalize it. The purpose of this short discussion is merely to give the reader some idea of what a Hamiltonian operator is and how to use it, to introduce some of the key words which will slip out later, and generally to sketch the framework which embraces our subsequent discussion.

III.2.3. Separation of electronic and nuclear motion

We are now ready to begin at the beginning. Most if not all readers will have a fairly clear idea of what a molecule is, and they will be familiar with the dynamic properties of molecules like electronic motion, vibrations, rotations, electron and nuclear spin effects, and so on. We are all used to treating these properties separately, according to our particular interests, but perhaps we do not always ask how these properties arise, or how far we are justified in partitioning the total energy in this manner. We know that it works rather well, but a good theory must explain why. If asked to describe in the most elementary and unprejudiced way what a molecule is, one might reply that it is a definite assembly of nuclei and electrons which are prepared to coexist in a certain configuration with considerable stability. The nuclei are positively charged, the electrons are negatively charged, and both types of particle can also possess the property of spin. We will return to the question of spin later, but let us accept it as an observed property for the present. A complete Hamiltonian for this assembly in the absence of external fields, $\mathcal{H}_{\text{total}}$, will consist of a sum of different types of terms as follows:

$\mathcal{H}_{\text{nucl}}$ representing the kinetic energy of the nuclei.
$\mathcal{H}_{\text{el}}$ representing the kinetic energy of the electrons.
V_{el} representing the potential energy due to electrostatic interactions between the electrons.
V_{nucl} representing the potential energy due to electrostatic interactions between the nuclei.

$V_{el, nucl}$ representing the potential energy due to electrostatic interactions between electrons and nuclei.

$\mathscr{H}(S)$ representing the magnetic interactions arising from the presence of electron spin S.

$\mathscr{H}(I)$ representing the magnetic and electric interactions arising from the presence of nuclear spin I, and also interactions between nuclear and electron spins.

Hence we may write,

$$\mathscr{H}_{total} = \mathscr{H}_{nucl} + \mathscr{H}_{el} + V_{el} + V_{nucl} + V_{el, nucl} + \mathscr{H}(S) + \mathscr{H}(I) \quad \text{(III.18)}$$

Now our aim is to find a way of rewriting this Hamiltonian as a sum of terms representing translational, electronic, vibrational, rotational, electron spin and nuclear spin effects respectively. There are, in fact, a number of ways of attempting to effect this partitioning of the energy, but there are always cross-terms between the different contributions. In other words, complete separation of the contributions is not possible, and at best we can only aim to minimize the cross-terms.

None of the terms in the Hamiltonian (III.18) looks particularly complicated at this stage. Since the kinetic energy of a particle of mass m moving with velocity $\mathbf{v}$ is equal to $\frac{1}{2}mv^2$, we can obviously write the first two terms of (III.18) as

$$\mathscr{H}_{nucl} + \mathscr{H}_{el} = \sum_{\alpha} \tfrac{1}{2}M_{\alpha}\mathbf{V}_{\alpha}^2 + \sum_{i} \tfrac{1}{2}m\mathbf{v}_i^2 \quad \text{(III.19)}$$

where M_α is the mass of nucleus α which is moving with velocity $\mathbf{V}_\alpha$ and m is the mass of each electron moving with velocity $\mathbf{v}_i$. Note, however, that we have written $\mathbf{V}_\alpha$ and $\mathbf{v}_i$ in bold-faced type to denote that they are vector quantities, that is, the velocity of a particle has both magnitude and direction. In order to measure a vector quantity we must define a coordinate system to act as a frame of reference, so we must start by defining, quite arbitrarily, an origin and cartesian coordinate system X, Y, Z as shown in Fig. III.1. We call this a "space-fixed axis system of arbitrary origin"—space-fixed in the sense that X, Y, Z are *not* to be identified with any axis system we might later define for the molecule itself. Figure III.1 illustrates a rather simple system consisting of two different nuclei with velocity vectors $\mathbf{V}_1$ and $\mathbf{V}_2$, and three electrons with velocity vectors $\mathbf{v}_1$, $\mathbf{v}_2$ and $\mathbf{v}_3$. Clearly we can describe the instantaneous position of each particle in terms of its appropriate position coordinates in the X, Y, Z axis system, and we can describe the velocity vectors in terms of components in the directions of X, Y, Z. Now consider what we would like to achieve, and think about the two nuclei first. We are probably all familiar with the idea that the total kinetic energy of the nuclei can be partitioned into

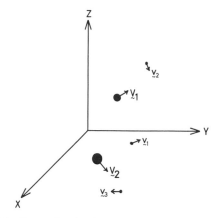

Fig. III.1. Space-fixed axis system with arbitrary origin.

three different types of motion. First, there is the translational motion, illustrated in Fig. III.2(a), in which the positions of the two nuclei are fixed relative to each other, but together they move through space in some direction (translational motion, of course, is normally taken to include the electrons as well). Second, there is the rotational motion, in which the two nuclei remain the same distance from each other, but rotate about their centre of mass, as shown in Fig. III.2(b). Third, there is the vibrational motion, in which the nuclei move along a line passing through them, either towards or away from each other, as shown in Fig. III.2(c).

Now it is not difficult to see that the motions of the nuclei depicted in Fig. III.1 can be decomposed into the sum of the three types of motion depicted in Fig. III.2, and the way in which this separation is achieved is through a transformation of either the origin or directions, or both, of the initial axis system described in Fig. III.1. We can say that we have *separated* a particular mode of motion if we can find a particular coordinate system in which it does not appear. The problem, then, is to find the appropriate coordinate systems.

The first coordinate transformation is to move the origin from the arbitrarily chosen position in Fig. III.1 to the centre of mass of all particles, nuclei and electrons, without changing the directions of the axes. Of course, since the particles are moving, the centre of mass is also moving, but this does not matter. If the origin of the coordinate system is always at the centre of mass, the translational motion of the whole molecule does not appear in the new coordinate system, and has therefore been separated. This coordinate transformation is illustrated in Fig. III.3. In practice the translational motion is completely separable in the absence of external fields; if external fields are

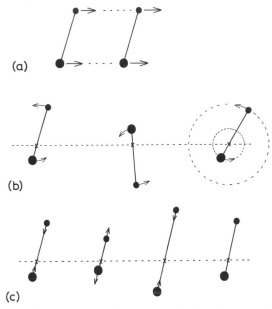

Fig. III.2. Motions of the nuclei in a diatomic molecule, (a) translation, (b) rotation, (c) vibration.

applied, however, cross-terms between the applied field and the translational motion appear, the effects of which are experimentally observable in some high-resolution molecular beam experiments. Because of the separation of the translational motion of the molecule as a whole, the velocity vectors of the individual particles will now look different in the new coordinate system. We have not, however, attempted to draw them!

Next we tackle the rotational motion of the *nuclei* by keeping the origin at the molecular centre of mass but changing the directions of the axes so that they rotate with the molecule. Actually a linear molecule (like our diatomic species) is an unfortunate example to illustrate this particular axis transformation, because although we may define the Z axis to lie in the same direction as a line joining the two nuclei, the X and Y axes are not uniquely determined. To be sure, they are both perpendicular to Z but any further choice is arbitrary. There are ways of getting round this problem in practice, which we need not go into here. If our molecule were a non-linear molecule like H_2CO, the directions of the new axes, which we call molecule-fixed axes, are fully determined, as shown in Fig. III.19. A common convention is to denote molecule-fixed axes by small letters x, y, z. Note that the translating axis system in Fig. III.3(b) can be related to an instantaneous position of the

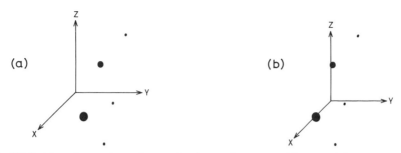

Fig. III.3. Transformation of space-fixed axes from an arbitrary origin to an origin at the centre of mass of all particles.

rotating molecule-fixed axes, through the use of the spherical polar angles θ and ϕ as shown in Fig. III.4. Hence the rotational motion of the nuclei, which is almost completely separated by this coordinate transformation, will be described by a Hamiltonian which contains the coordinates θ and ϕ.

Finally, we can largely separate the vibrational motion of the two nuclei by moving the origin from the molecular centre of mass (which includes contributions from the electrons) to the nuclear centre of mass. If we always define the coordinate origin to be at the nuclear centre of mass, the vibrational motion of the nuclei is largely separated from the other degrees of freedom in this coordinate system.

Each transformation described above yields a Hamiltonian operator describing the motion which is partitioned off by the transformation. The Hamiltonian describing the translational motion of the whole molecule is just

$$\mathcal{H}_{\text{trans}} = \frac{\mathbf{P}^2}{2M} \tag{III.20}$$

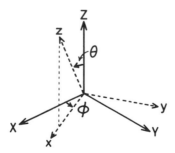

Fig. III.4. Definition of the spherical polar angles θ and ϕ relating the space-fixed axis system to the rotating axis system.

where **P** and M are the momentum and total mass respectively of the molecule. Apart from its interaction with applied fields and its effect on line widths, this contribution to the total energy is relatively uninteresting to most spectroscopists.

The rotational kinetic energy of the nuclei obtained after the second transformation takes the form

$$\mathcal{H}_{\text{rot ke}} = -\frac{\hbar^2}{2\mu r^2} \left\{ \operatorname{cosec} \theta \frac{\partial}{\partial\theta} \left(\sin \theta \frac{\partial}{\partial\theta} \right) + \operatorname{cosec}^2 \theta \frac{\partial^2}{\partial\phi^2} \right\} = B\mathbf{R}_2$$

(III.21)

where B is called the rotational constant, **R** is the rotational angular momentum, and r is the internuclear distance.

The vibrational kinetic energy of the nuclei takes the form

$$\mathcal{H}_{\text{vib ke}} = -\frac{\hbar^2}{2\mu r^2} \left\{ \frac{\partial}{\partial r} \left(r^2 \frac{\partial}{\partial r} \right) \right\}.$$

(III.22)

Now we can compare equations (III.20), (III.21) and (III.22) with the Hamiltonian (III.18) from which we started. Setting aside the spin terms for the moment, we might have hoped to achieve through our coordinate transformations a separation of the total Hamiltonian into a part describing the translational motion of the whole molecule, a part describing the nuclear kinetic energy, and a part describing the motion of the electrons relative to a *fixed* nuclear configuration. Unfortunately such a complete separation has not been realized; the spanner in the works is the potential energy term $V_{\text{el, nucl}}$ which describes the electrostatic interactions between the electrons and nuclei, and clearly varies as the nuclei move. The task is thus to attempt to remove the effects of nuclear motion from the electronic part of the Hamiltonian.

The first coordinate transformation described above was carried out for all particles so as to separate the translational motion of the whole molecule, but the two subsequent transformations were described for the nuclei alone. The obvious method of eliminating the dependence of the electronic coordinates on nuclear motion is to transform the electronic variables (spatial and spin) from the space-fixed coordinate system to the rotating axis system. However this transformation has subtle effects on the rotational Hamiltonian (III.21). A tricky point to appreciate is that the rotational Hamiltonian derived actually describes the rotational motion of the nuclei in the *space-fixed axis system*; the purpose of the transformation to rotating axes was to separate the nuclear motion and to yield an expression giving the nuclear kinetic energy in terms of angles describing the transformation.

Hence after the first transformation we are left with the electron position (space), electron spin, and nuclear spin coordinates expressed in a space-fixed axis system with origin at the molecular centre of mass. A transformation to the rotating axis system can be applied independently to any one of these coordinates, or to all of them if we wish. It is possible, therefore, to distinguish several possible combinations of transformations and these turn out to be closely related to Hund's coupling cases for diatomic molecules which we describe in detail in section III.4. Case (a) corresponds to transformation of both the electron position and spin coordinates to the rotating axis system, whilst case (b) involves transformation of the electron position coordinates only. It does not matter much whether we choose to transform the nuclear spin coordinates or not. When we carry out these transformations we find that the nuclear rotational Hamiltonian can be rewritten; in the case (a) transformation the rotational Hamiltonian changes according to

$$\mathcal{H}_{rot} = B\mathbf{R}_2 = B\{\mathbf{J} - \mathbf{L} - \mathbf{S}\}^2 \tag{III.23}$$

where $\mathbf{S}$ is the electron spin angular momentum, $\mathbf{L}$ is the orbital angular momentum and $\mathbf{J}$ is the total angular momentum of the molecule. $\mathbf{L}$, in fact, describes how the electrons move with respect to the two nuclei. In the case (b) transformation the rotational Hamiltonian becomes

$$\mathcal{H}_{rot} = B\mathbf{R}_2 = B\{\mathbf{N} - \mathbf{L}\}^2 \tag{III.24}$$

where $\mathbf{N}$ is the total angular momentum *except* for the electron spin. We shall appreciate the significance of (III.23) and (III.24) as our discussion progresses. Although we have now removed the dependence of $V_{el, nucl}$ on the rotational coordinates, we have not separated electronic and nuclear motion because equations (III.23) and (III.24) involve the electron orbital motion as well as the nuclear rotational motion.

We recall that the purpose of the coordinate transformations was to produce, as far as possible, expressions describing the partitioning of the total energy of the molecule into translational, vibrational, rotational, electron space, electron spin and nuclear spin parts. We have discussed in detail the first three parts and in the following sections we outline the remaining parts of the Hamiltonian. Although we have here dealt with diatomic molecules, the principles apply also to non-linear polyatomic molecules. The details however, will be discussed later when we come to deal with particular molecules.

III.2.4. Dirac theory of electron spin

The notion that the electron possesses an intrinsic angular momentum, which is called spin, was first introduced by Pauli to account for unexpected splittings in the optical spectra of alkali metals. A further convincing experimental demonstration of this intrinsic angular momentum and a magnetic moment associated with it was provided by Stern and Gerlach (1924). They passed a beam of silver atoms through an inhomogeneous magnetic field and observed that the beam split into two spatially separated components. It was well known that the electron possessed a magnetic moment due to its orbital motion, and the additional postulate of a spin angular momentum $\mathbf{S} = \frac{1}{2}$, so that two spatial orientations of the spin are possible, was found to account for many otherwise unexpected observations. The spin magnetic moment seemed, however, to be twice as large as expected and was therefore termed an "anomalous" magnetic moment.

Dirac (1928) was the first person to provide a convincing theoretical explanation of this new property. He set out to derive a wave equation for the electron which satisfactorily obeyed the laws of relativity, introduced earlier by Einstein. The fact that some of the electrons in atoms move with a velocity approaching that of light suggests that a relativistic description of electrons should be more accurate. Through a remarkable process of reasoning, Dirac decided that the correct wave equation must be of the form

$$\{E - c\boldsymbol{\alpha}.\mathbf{P} - \beta mc^2\}\psi = 0 \tag{III.25}$$

(corresponding to a Hamiltonian, $\mathscr{H} = c\boldsymbol{\alpha}.\mathbf{P} + \beta mc^2$) where the eigenfunctions ψ have energies E, m is the mass of the electron, c is the velocity of light and $\mathbf{P}$ is the momentum of the electron. The quantities $\boldsymbol{\alpha}$ and β are unusual, however, in that they are represented by four 4×4 matrices, three corresponding to the components of the vector $\boldsymbol{\alpha}$, and one to β. Thus (III.25) is actually a four-dimensional wave equation and the function ψ is therefore a four-component wave function. On solving his wave equation, Dirac found that two types of solution corresponded to states of positive energy, as required for an electron, but the other two types corresponded to states of negative energy. He was thus led to postulate the existence of the positron, a new particle which was subsequently found by experimental particle physicists.

Hence it is desirable to reduce equation (III.25) to one which involves only the electron, and is therefore two-dimensional. Any such reduction must, however, still obey the laws of relativity and therefore include the effects of coupling between the positive and negative energy states. Several reduction methods have been described, but probably the best known is a transformation due to Foldy and Wouthuysen (1950). This is not the place to describe the details, although they are not difficult and the reader who wants to understand fully this fascinating subject is advised to consult the original papers.

The conclusions are extremely important, however, and they are as follows:

(i) Examination of the electron position coordinate shows that the electron exhibits a periodic oscillatory spiralling motion, called the Zitterbewegung by Schrodinger. This spiralling motion of the charged particle gives rise to the magnetic moment; the usual concept of a spinning motion about an axis through the particle may be helpful, but is essentially bogus!

(ii) What we call the spin angular momentum S has the value $\frac{1}{2}$, and two orientations of the spin are allowed. These two orientations correspond to the two components of the wave function for the electron. (Similarly, the positron also has two spin components.)

(iii) The Dirac equation for an electron in an electromagnetic field can also be reduced to a two-dimensional wave equation, and one finds that the magnetic moment due to the spin is $g_e\mu_B S$ where g_e has the value 2, rather than 1 as for the corresponding orbital magnetic moment. The "anomalous" magnetic moment is thus accounted for.

The electron and nuclear spin Hamiltonians used in interpreting microwave spectra contain a number of terms representing interactions between pairs of spins, either spin being electronic or nuclear. These terms may be derived theoretically from the Dirac equation for the electron by introducing the magnetic fields due to the other particles in a purely classical manner. This may seem somewhat artificial, but it does at least provide a systematic method of obtaining the various spin-spin interaction terms. Alternatively one may use an equation due to Breit as the starting point. This equation consists essentially of two Dirac equations, one for each particle, plus a term representing coupling between the particles. The Dirac and Breit equations actually describe particles with spin $\frac{1}{2}$, and the effects of nuclei with spins greater than $\frac{1}{2}$ are incorporated empirically. It should be said, however, that even for spin $\frac{1}{2}$ particles a more rigorous derivation of some of the terms appearing in, for example, an effective nuclear spin Hamiltonian has yet to be developed.

III.2.5. Nuclear hyperfine interactions

There are a number of ways in which a nucleus can have magnetic or electrostatic interactions with the electrons in a molecule. These interactions produce nuclear hyperfine splittings in microwave spectra, and although these splittings are often very small, their observation is important since they provide valuable information about the relative positions of the nuclei and electrons. In other words, measurements of nuclear hyperfine effects provide excellent tests of computed molecular wave functions.

Not all nuclei exhibit hyperfine effects, but only those which possess a nuclear spin angular momentum **I**. We shall not go into the details of what determines the magnitude of the spin for a given nucleus, since this is very much part of nuclear physics. Molecular spectroscopists are usually prepared to accept that a given nucleus is known to have a certain spin **I** and an associated magnetic moment $\boldsymbol{\mu}_I$. The magnetic moment is, in fact, proportional to the spin according to the equation

$$\boldsymbol{\mu}_I = g_I \mu_N \mathbf{I} \tag{III.26}$$

where μ_N is the nuclear magneton and g_I, the nuclear g factor, has a characteristic and different value for each type of nucleus. Since the spin and orbital motion of the electrons in a molecule also give rise to associated magnetic moments, it is not difficult to appreciate the likelihood of magnetic coupling between the nuclear and electronic magnetic moments. The various ways in which this can occur are described in this section.

Nuclei also possess positive electrostatic charges and since the electrons are negatively charged, it is again easy to appreciate the possibility of electrostatic interactions between the nuclei and electrons. We shall see that after the monopolar interactions, the next most important electrostatic interaction is the so-called "nuclear electric quadrupole interaction", which arises because the nuclei have a finite size. Only nuclei possessing net spin angular momentum greater than $\frac{1}{2}$ also possess the electrostatic charge distribution necessary for quadrupole interactions to occur; this problem is also discussed in detail later.

(a) *Nuclear-electron spin dipolar interaction*

When we accept it as obvious that the magnetic moments due to nuclear and electron spin are likely to interact, we are really visualizing the dipole–dipole interaction which has a direct classical analogy in the interaction between two bar magnets. We deal with this interaction first because it is important, but also because it enables us to introduce anisotropy, which is an important feature of many intramolecular interactions.

Let us first suppose that the electron and nuclear magnetic moments can be regarded as point dipoles separated by a distance r. We define a cartesian axis system x, y, z and now consider the dipole coupling when both dipoles point in the direction of the z axis. This would, for example, be the situation when a very strong magnetic field is applied in the z direction. We take the origin of the cartesian coordinate system to be at the position of the electron. The line joining the particles is a vector **r** since it has a definite direction relative to the cartesian axes, defined by means of the polar angles θ and ϕ (see Fig. III.5).

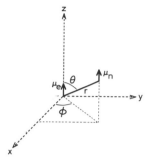

Fig. III.5. Coordinate system for the dipolar interaction between electron and nuclear spin magnetic moments.

Now classical magnetostatics tells us that the interaction energy of two classical dipoles $\boldsymbol{\mu}_e$ and $\boldsymbol{\mu}_I$ is given by

$$E = \frac{\boldsymbol{\mu}_e \cdot \boldsymbol{\mu}_I}{r^3} - \frac{3(\boldsymbol{\mu}_e \cdot \mathbf{r})(\boldsymbol{\mu}_I \cdot \mathbf{r})}{r^5}. \tag{III.28}$$

We can apply this equation to our problem by noting the standard result that the vector $\mathbf{r}$ can be written as the sum of components in the x, y, z directions, each component consisting of a unit vector ($\mathbf{i}$, $\mathbf{j}$, $\mathbf{k}$ in the x, y, z directions) multiplied by the appropriate magnitude of the component r_x, r_y or r_z. The magnetic moment vectors $\boldsymbol{\mu}_e$ and $\boldsymbol{\mu}_I$ can likewise be written in terms of components in the directions of the axes, and since in our example we take the dipoles to point in the z direction, the x and y components are zero. Hence the expression for the dipolar energy becomes

$$E = \frac{\mu_e \mu_I \mathbf{k}^2}{r^3} - \frac{3(\mu_e r_z)(\mu_I r_z)\mathbf{k}^2}{r^5} \tag{III.29}$$

$$= \frac{\mu_e \mu_I}{r^3}\left\{1 - 3\frac{r_z^2}{r^2}\right\}. \tag{III.30}$$

Now the cartesian components r_x, r_y, r_z can be written in terms of the polar coordinates so that

$$r_x = r \sin\theta \cos\phi, \qquad r_y = r \sin\theta \sin\phi, \qquad r_z = r \cos\theta \tag{III.31}$$

Substituting for r_z in (III.30) we obtain the result for the dipolar interaction energy,

$$E = \frac{\mu_e \mu_I}{r^3}\{1 - 3\cos^2\theta\} \tag{III.32}$$

The important feature of this result is, of course, the angular dependence. The measured interaction energy depends upon the angle θ between the line joining the dipoles and the direction of the field which orients them. This angular dependence is called anisotropy. Our example is a simple one because in other cases the magnetic moments $\boldsymbol{\mu}_e$ and $\boldsymbol{\mu}_I$ may not be parallel to each other.

In a molecular system this classical description of the dipolar interaction between electron and nuclear spin magnetic moments requires some modification. Firstly the classical moments $\boldsymbol{\mu}_e$ and $\boldsymbol{\mu}_I$ must be replaced by the corresponding quantum mechanical operators,

$$\boldsymbol{\mu}_e = g_e\mu_\mathrm{B}\mathbf{S}, \qquad \boldsymbol{\mu}_I = g_I\mu_\mathrm{N}\mathbf{I}. \tag{III.33}$$

Secondly, although it may be legitimate to regard the spinning nucleus as a localized point dipole, this description certainly cannot apply to the electron. The "position" of an electron is described by considering the electronic wave function and calculating the probability distribution in different regions of space. If, for example, it were accurate in a particular case to say that the unpaired electron occupied an atomic p orbital, the dipolar coupling to a nucleus would be computed by integrating over all possible positions of the electron in that orbital. Hence measurement of the dipolar coupling provides a sensitive test of the electronic wave function.

For free radicals in solids, where the orbital angular momentum is largely removed by the crystalline environment, it is usually true that the electron and nuclear spins are strongly quantized by an applied magnetic field. In other words, the magnetic moments are aligned in the direction of the field. For gaseous molecules with orbital angular momentum, however, the situation is quite different. The electron spin is then strongly aligned by its coupling to the orbital angular momentum, and is not appreciably decoupled by an applied field. For NO in its ground state, for example, we can fairly accurately describe the magnetic moments due to $\mathbf{L}$ and $\mathbf{S}$ as lying in the direction of the internuclear axis. Consequently any hyperfine coupling component in a direction perpendicular to the axis is very small. The reader may well wonder how it is possible to measure the component of hyperfine interaction in a particular direction when the molecule is rotating in the gas phase. The answer is that although the molecule is rotating it is in a well defined quantum state; in liquids, on the other hand, the rotational motion is not quantized because of frequent molecular collisions which lead to randomization. In the latter circumstances the dipolar coupling becomes a wholly time dependent random property whose average value is zero. Hence the dipole–dipole coupling can be measured in gases and solids, but not in normal liquids.

Because the dipole–dipole coupling is anisotropic, it is usually represented in an effective Hamiltonian by a term of the form

$$\mathbf{S} . \mathbf{T} . \mathbf{I} \tag{III.34}$$

$\mathbf{T}$ is called a second-rank tensor and is represented by a 3×3 matrix. In an arbitrary cartesian coordinate system x, y, z, the tensor $\mathbf{T}$ has nine components,

$$\begin{vmatrix} T_{xx} & T_{xy} & T_{xz} \\ T_{yx} & T_{yy} & T_{yz} \\ T_{zx} & T_{zy} & T_{zz} \end{vmatrix} \tag{III.35}$$

which are all, in general, non-zero. However when we describe intramolecular properties we usually choose a coordinate system which is closely related to the geometry of the molecule. Now it is always possible to choose a coordinate system which reduces the coupling tensor $\mathbf{T}$ to diagonal form, that is, renders the off-diagonal elements zero. This axis system will usually be one related to the symmetry of that part of the molecule where the dipolar coupling is important. In linear molecules, for example, it is customary to define the z axis to be coincident with the molecular axis; for non-linear molecules in the gas phase, the experimental measurements are usually interpreted in terms of axes which are coincident with the inertial axes of the molecule.

(b) Fermi contact interaction

A convincing physical description of the Fermi contact interaction is difficult to provide, although its theoretical derivation from quantum mechanics is relatively straightforward. It was first introduced by Fermi to account for some unexpected splittings in the optical spectra of atoms, and requires that the unpaired electron have a non-zero density at the nucleus. Hence the term "contact interaction". The requirement of a non-zero electron spin density at the nucleus means, of course, that the unpaired electron or electrons must in part occupy an atomic s orbital. In contrast to the dipolar coupling, the Fermi contact interaction is an isotropic interaction. Consequently it is readily observed in gases, liquids and solids, and in an effective Hamiltonian is always represented by a term of the form

$$a\mathbf{S} . \mathbf{I} \tag{III.36}$$

where a is the isotropic coupling constant. The magnitude of a varies from one type of s orbital to another, and from one atom to another. It is also worth noting that the contact interaction is nearly always observable, even in those cases when a simple model of the electronic wave function would suggest that s orbitals are not involved. The fact is that simple models are never very accurate, and the spin density at the nucleus is nearly always finite.

Although it may seem obvious, it is nevertheless worth remarking that the contact interaction and the dipolar interaction are usually both present; provided one can study the angular dependence of the total hyperfine splitting, either by studying orientational dependence in the solid state, or rotational dependence in the gas phase, it is usually possible to separate the isotropic and anisotropic parts of the hyperfine coupling.

(c) *Nuclear spin–electron orbital hyperfine coupling*
In molecules where there is a net orbital angular momentum, there is a dipolar coupling of the orbital magnetic moment with the nuclear spin magnetic moment, analogous to the electron spin-nuclear spin dipolar coupling described earlier. This interaction is especially important in linear molecules like NO and NCO which have $^2\Pi$ electronic ground states. It is relatively unimportant in most non-linear molecules, however, which do not normally possess quantized orbital angular momentum. It is also usually unimportant in condensed phases, where the orbital motion is quenched by the environment.

(d) *Nuclear electrostatic quadrupole interaction*
Nuclei carry a net positive electrostatic charge but this charge is not necessarily distributed evenly; the charge distribution can, however, be decomposed into the sum of different symmetrical charge distributions corresponding to monopole, quadrupole, hexadecapole moments, etc. In an atom or molecule the charged nucleus interacts with the electrostatic fields due to the electrons and other nuclei. The electrostatic nucleus-nucleus and nucleus-electron monopole-monopole interactions make important contributions to the total energy of the molecule, and they are represented by terms which appear in the electronic and vibrational Hamiltonians. We are here concerned with the interaction between the nuclear quadrupole moment and an electric field *gradient* produced primarily by the surrounding electrons. Although this interaction makes only a small contribution to the total energy of the molecule, it produces splittings in microwave spectra which yield valuable information about the electronic structure.

In order to understand the quadrupole interaction we must first consider the allowed spatial orientations of a nucleus of spin **I**. Suppose we define a space-fixed direction Z as being the direction of an applied magnetic field. The nuclear magnetic dipole moment will interact with the field to orient the nucleus in space, but only a certain number of specific orientations are allowed by quantum theory. If the value of the spin is I, there are $2I + 1$ allowed orientations with respect to Z, the component of the spin in the Z direction being characterized by a quantum number M_I which takes the values $I, I - 1, I - 2, \ldots, -I$. Hence for a nucleus of spin $\frac{1}{2}$, such as the proton, there are just two orientations allowed, corresponding to $M_I = +\frac{1}{2}$ or $-\frac{1}{2}$. For the ^{35}Cl nucleus, with $I = \frac{3}{2}$, there are four allowed orientations corresponding to $M_I = +\frac{3}{2}, +\frac{1}{2}, -\frac{1}{2}, -\frac{3}{2}$. In a magnetic field these four orientations have energies proportional to the corresponding value of M_I.

Now in the case of the electrostatic quadrupole interaction, we enquire about the interaction between an electrostatic field gradient (which now defines our Z direction) and the nuclear quadrupole moment. As we mentioned in the first paragraph of this section, the total charge distribution of the nucleus may be decomposed into the sum of monopole, quadrupole, hexadecapole moments, etc. and the quadrupolar distribution might, for example, be represented as a cigar shaped distribution having cylindrical symmetry about a principal axis fixed in the nucleus, which we define as the nuclear z axis. That this corresponds to a quadrupolar distribution of charge may be readily appreciated by considering the nuclear charge distribution at symmetrically disposed points on the $+z, -z, +x, -x$ axes. The nuclear charge at the $\pm x$ points is clearly $\delta -$ with respect to that at the $\pm z$ points (see Fig. III.6). Now for a nucleus of spin $I = 1$ there are three allowed spatial orientations of the spin. Taking the space-fixed Z axis as the axis of quantization, these three orientations may be identified with those in which the nuclear z axis is coincident with Z (Fig. III.6(a)), perpendicular to Z (Fig. III.6(b)), and antiparallel to Z (Fig. III.6(c)). These three orientations correspond to $M_I = +1, 0, -1$ respectively. If now the nucleus is in an electrostatic field

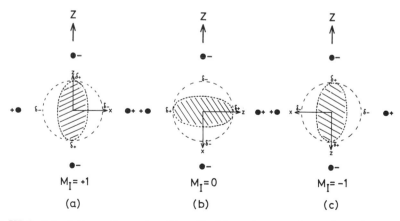

Fig. III.6. Orientations of a nucleus ($I = 1$) with an electrostatic quadrupole moment in an electrostatic field gradient.

gradient due to the electrons (as shown, for example, by the point charges in Fig. III.6) it is clear that the state of $M_I = 0$ has a different electrostatic energy from the states of $M_I = \pm 1$. There is therefore an electrostatic quadrupole splitting of the nuclear states whose magnitude depends upon the sizes of the nuclear quadrupole moment and the electrostatic field gradient.

If a magnetic field is also applied, the magnetic interaction with the nuclear magnetic dipole moment also acts to orient the spin and the final result is a balance between the electrostatic and magnetic interactions. Nuclear magnetic hyperfine interactions may also compete with both the quadrupole and external field interactions.

III.2.6. Interactions with external fields

Spectroscopic experiments consist of perturbing molecules with external fields and studying their response. Spectroscopic transitions between energy levels are induced by allowing a molecule to interact with electromagnetic radiation, which consists of oscillating magnetic and electric fields. In many experiments static fields are also applied, as we have seen in Chapter II. In this section, therefore, we consider in more detail how molecules can interact with applied magnetic or electric fields, and what the consequences are.

We consider first the application of static magnetic fields, which are particularly important in the study of open shell molecules. A molecule interacts with an applied magnetic field ($\mathbf{B}$) by virtue of its magnetic moment, but the latter arises from a number of different sources which must be considered individually. The most important of these in open shell molecules is electron spin. The magnetic dipole moment $\boldsymbol{\mu}_e$ arising from electron spin is given by

$$\boldsymbol{\mu}_e = g_e \mu_B \mathbf{S} \tag{III.37}$$

where $\mathbf{S} = \frac{1}{2}$, μ_B is the Bohr magneton ($\mu_B = e\hbar/2mc$) and g_e is a quantity which we shall call the "free electron spin g factor". As already discussed in section III.2.4, the Dirac theory predicts $g_e = 2$, if the applied field is treated classically. In fact one finds the free electron g value to be very close to $2 \cdot 0023$, and this value is accurately accounted for by treating the applied field quantum mechanically (Schwinger, 1948, 1949), rather than classically. In addition, there is a further small relativistic correction due to the high electron velocity in the molecule.

The next important interaction to be considered is that between the applied field and the magnetic moment arising from electronic orbital angular momentum $\mathbf{L}$, the latter being given by

$$\boldsymbol{\mu}_L = g_L \mu_B \mathbf{L} \tag{III.38}$$

g_L is the orbital g factor and has the value unity, except for possible very small relativistic corrections. There is, incidentally, no quantum field theory correction to g_L analogous to that described above for g_e. Hence the combined effect of interaction of $\mathbf{B}$ with $\mathbf{S}$ and $\mathbf{L}$ is to yield two terms in the Hamiltonian, which are

$$\mathscr{H} = g_e\mu_B\mathbf{B}.\mathbf{S} + g_L\mu_B\mathbf{B}.\mathbf{L}. \tag{III.39}$$

Whether or not these terms appear in the "effective Hamiltonian" depends upon the particular molecule under investigation. For nitric oxide, where both $\mathbf{S}$ and $\mathbf{L}$ are quantized, the electronic Zeeman effect in the $^2\Pi_{\frac{1}{2}}$ state would be represented by (III.39) in the effective Hamiltonian. If, however, we are dealing with an electronic state in which a component of $\mathbf{L}$ is not preserved (for example, almost any electronic state in a non-linear molecule), the second term in (III.39) does not usually appear explicitly because it does not make a first-order contribution to the energy. (This term does not appear explicitly either for a Σ state in a linear molecule.) However it can make a second-order contribution, generating some orbital angular momentum by mixing excited electronic states with the ground state. This effect is discussed further in the following section.

The reader who is familiar with solid state electron spin resonance will also have encountered the idea of the g tensor; in this formalism the interaction of the applied field with the electron spin is represented by a term of the form

$$\mathscr{H} = \mu_B\mathbf{B}.\mathbf{g}.\mathbf{S}' \tag{III.40}$$

where $\mathbf{g}$ is usually regarded as a second rank tensor which has three principal values (often different) corresponding to $\mathbf{B}$ being applied in three different orthogonal directions defined in the molecule. Two points should be noted here. First, equation (III.40) is an effective spin Zeeman interaction term which represents the combined effects of both terms in equation (III.39). Second, the spin $\mathbf{S}'$ in (III.40) is an effective (often called fictitious) spin and is not, in general, the same quantity as the true spin $\mathbf{S}$ appearing in (III.39). We will not pursue this aspect further, however, because the concept of fictitious spin is not usually employed in the analysis of gas phase spectra.

Two further interactions arising from an external magnetic field remain to be considered. First there is the interaction with the rotational magnetic moment of the nuclei, the interaction being represented in a diatomic molecule by the term

$$\mathscr{H} = g_r^n\mu_B\mathbf{B}\cdot\mathbf{R} \tag{III.41}$$

where g_r^n is the nuclear rotational g factor and $\mathbf{R}$ is the rotational angular momentum of the nuclei in the case of a diatomic molecule. For more complicated molecules there are, of course, further terms in the rotational Zeeman interaction. Now as we shall see later, the rotational magnetic moment of a molecule contains a contribution from the electrons as well as from the nuclei. The electronic contribution arises from a cross-term between the orbital Zeeman interaction and the rotational Hamiltonian; consequently in a diatomic molecule the total rotational Zeeman effect is more often represented by a term of the form

$$\mathcal{H} = -g_r\mu_B\mathbf{B}.\mathbf{J} \qquad \text{(III.42)}$$

where $\mathbf{J}$ is the total angular momentum of the nuclei and electrons and g_r is the difference between a contribution g_r^e from the electrons and a contribution g_r^n from the nuclei.

Finally we must include the nuclear spin Zeeman interaction

$$\mathcal{H} = -g_I\mu_N\mathbf{B}.\mathbf{I} \qquad \text{(III.43)}$$

g_I is the nuclear g factor and has a characteristic value for each type of nucleus, μ_N is the nuclear magneton and $\mathbf{I}$ is the nuclear spin. The direct effects of this term are straightforward, but once again it can combine with other terms to give more subtle second-order effects, some of which we shall discuss in the next section.

Turning now to the effects arising from application of a static electric field, the position at first sight seems to be a great deal simpler. A uniform electric field $\mathbf{E}$ can only interact with the molecular electric dipole moment μ, and the interaction, called the Stark interaction, is represented by the term

$$\mathcal{H} = -\mu.\mathbf{E}. \qquad \text{(III.44)}$$

This apparently simple term leads, however, to considerable complications when it is examined in more detail. The electric field direction is naturally described in terms of a set of space (laboratory)-fixed coordinates, whereas the dipole moment of the molecule, which depends upon the distribution of nuclear and electronic charge, is defined in terms of molecule-fixed axes rotating with the molecule. Consequently in order to expand the scalar product of μ and $\mathbf{E}$ in equation (III.44) we must either express $\mathbf{E}$ in terms of components along the molecule-fixed axes, or alternatively express the components of μ in terms of space-fixed axes. Whichever method we choose, the rewritten Stark Hamiltonian will contain terms relating the space- and molecule-fixed axes, and consequently will operate on rotational basis functions in a fairly complicated manner. (This is equally true for the Zeeman Hamiltonian.) We shall be giving examples of the Stark effect in Chapters IV and V.

So much for static fields, for the present at least. Electromagnetic radiation of any frequency can be regarded as a combination of oscillating magnetic and electric fields perpendicular to each other and also perpendicular to the direction of propagation of the radiation. The oscillating magnetic component will interact with the magnetic dipole moment of the molecule, whilst the oscillating electric field interacts with the molecular electric dipole moment. Consequently the Hamiltonian describing the interaction of a molecule with electromagnetic radiation contains terms identical to those outlined for static fields, except that the static fields **B** and **E** are replaced by time-dependent fields $\mathbf{B}(t)$ and $\mathbf{E}(t)$ whose amplitudes oscillate periodically at a characteristic frequency. As we saw in the first chapter, transitions between two states of the molecule are induced if they are coupled by either the magnetic or electric components of the radiation; the transitions are therefore said to be either magnetic dipole or electric dipole allowed, and by examining whether the states are so coupled (i.e. evaluating the matrix element in equation (I.3)) we arrive at the spectroscopic selection rules and expressions for the relative intensities of different transitions. We reserve the details until Chapters IV and V, where we examine many different examples.

III.2.7. The construction of effective Hamiltonians

The reader who has persevered to this point might well be feeling that he (or she) too has acquired some angular momentum! Let us therefore attempt to summarize rather more simply some of the aspects covered, and at the same time introduce some further ideas. The interpretation of the microwave spectra of linear and non-linear radicals involves understanding how the different sorts of angular momenta interact with each other to produce the energy levels studied. The important vector quantities are **R**, the rotational angular momentum of the nuclei; **L**, the electronic orbital angular momentum; **S**, the electron spin angular momentum; **I**, the nuclear spin angular momentum; **B**, an applied magnetic field. We illustrate these vector quantities and the ways in which they can interact in Figs. III.7 and III.8; we see that each angular momentum can interact with the other three in turn, and in particular Fig. III.7 draws attention to the following pairwise interactions:

(i) **(L)(S)** —the spin-orbit coupling.

(ii) **(L)(R)** —rotational-electronic interaction.

(iii) **(L)(I)** —nuclear hyperfine interaction between the nuclear spin and electronic orbital magnetic moments.

(iv) **(S)(I)** —nuclear hyperfine interaction between the nuclear and electron spin magnetic moments.

(v) **(S)(R)** —magnetic coupling of the electron spin magnetic moment with the rotational magnetic moment.

(vi) **(R)(I)** —magnetic coupling of the nuclear spin magnetic moment with the rotational magnetic moment.

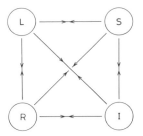

Fig. III.7. Possible pairwise interactions of **L**, **S**, **R** and **I**.

We can illustrate the interactions with an external magnetic field in a similar way, as shown in Fig. III.8; the four Zeeman couplings shown are as follows:

(i) **(B)(S)**—interaction between the applied field and the electron spin magnetic moment.

(ii) **(B)(I)** —interaction between the applied field and the nuclear spin magnetic moment.

(iii) **(B)(L)**—interaction between the applied field and the orbital magnetic moment.

(iv) **(B)(R)**—interaction between the applied field and the rotational magnetic moment.

Now all the terms illustrated in Figs. III.7 and III.8 appear in the complete Hamiltonian and they may well appear in the same form in an appropriate effective Hamiltonian constructed to analyse a particular spectrum. However it is also possible for the effective Hamiltonian to contain terms which account for features of the spectrum, but which do not appear in the same form in the complete Hamiltonian. This is best illustrated with some examples, two of which are shown in Fig. III.9. We recall that a direct interaction between **R**

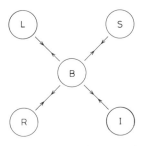

Fig. III.8. Possible interactions of an applied magnetic field **(B)** with **L**, **S**, **I** and **R**.

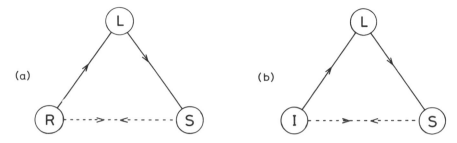

Fig. III.9. (a) Second-order spin-rotation interaction occurring *via* L. (b) Second-order pseudo-contact hyperfine interaction occurring *via* L.

and S can occur, but Fig. III.9(a) illustrates how an indirect coupling can also take place. The rotational angular momentum **R** interacts with **L**, which in turn couples with the spin **S**. Consequently the effective Hamiltonian may contain a term of the form $(\mathbf{R})(\mathbf{S})$, part of which arises from the direct coupling shown in Fig. III.7, but the remaining part coming from the indirect coupling via **L**. Now in a linear molecule where the electrons are free to execute orbital motion about the molecular axis, both terms may appear explicitly in the effective Hamiltonian. In a non-linear molecule, however, the lower symmetry restricts the orbital angular momentum and **L** does not then appear explicitly in the effective Hamiltonian. The residual effect of spin-orbit coupling, however, is to mix the ground electronic state with one or more excited electronic states; hence an observed apparent spin-rotation coupling arises partly by direct coupling, but partly through this admixture of excited states which generates some orbital angular momentum (Van Vleck, 1951). Clearly, the molecular constants associated with the spin-rotation term in the effective Hamiltonian must be carefully interpreted. In fact we shall see that for most non-linear molecules, the indirect contribution actually dominates the direct effect. The same considerations apply to linear molecules, like O_2, where there is no net orbital angular momentum in the pure $^3\Sigma$ ground electronic state.

Figure III.9(b) shows a similar situation with the nuclear hyperfine interaction. There is a direct interaction between the electron and nuclear spin magnetic moments, but an indirect interaction via the spin-orbit mixing of excited electronic states with the ground state can also occur. This is called the pseudo-contact interaction; usually it is rather small compared with the direct interaction. There is also the possibility of an indirect nuclear rotation–nuclear spin interaction occurring via **L**, which we have not illustrated.

Similar situations arise when we have, say, two electrons with unpaired spins (S_1 and S_2) and two nuclear spins, I_1 and I_2. Figure III.10(a) shows the couplings which can occur, either directly, or indirectly through the effects of L. Consider first the coupling of the two electron spins. There is certainly a direct interaction between the magnetic moments of the two electron spins, but there is also an indirect interaction which takes place via the spin-orbit mixing of excited and ground electronic states through the product of terms of the form $\{(S_1)(L)\}\{(L)(S_2)\}$. A good example is molecular oxygen, where the indirect coupling is appreciably larger than the direct interaction. With the related but heavier molecules SO and SeO, in which the spin-orbit coupling is even more important, the indirect coupling almost swamps the direct interaction. However, so far as the interpretation of a microwave spectrum is concerned, the effective Hamiltonian will usually contain a term of the form $(S_1)(S_2)$, and L will not appear explicitly, even though it is all important. So once more, the molecular constants which describe the so-called "spin-spin interaction" are not what they may seem to be. Very much the same arguments apply to the apparent $(I_1)(I_2)$ coupling in Fig. III.10(a), which can occur directly, or indirectly through coupling to the electron orbital angular momentum, that is, through terms of the form $\{(I_1)(L)\}\{(L)(I_2)\}$.

Fig. III.10. (a) Second-order spin-spin interactions occurring *via* L. (b) Electron coupled nuclear spin-spin interaction.

The reader will note that, so far, the indirect interactions we have described all involve the orbital angular momentum L, and indeed this is commonly true. Nevertheless there are other indirect interactions which do not involve L and one of these is illustrated in Fig. III.10(b). This is the so-called "electron coupled nuclear spin–spin interaction". The nuclear spin I_1 interacts with electron spin S_1; S_1 is coupled with S_2, which in turn interacts with I_2. The effect on the spectrum is that of an apparent direct nuclear spin–nuclear spin interaction, and the effective Hamiltonian will contain an appropriate term of that type. As in our previous examples, however, the coupling actually involves mixing of excited electronic states with the ground state, and it is not necessary for the total spin $S = S_1 + S_2$ to be non-zero. In other words, this indirect coupling can occur in closed-shell molecules, and indeed it is the origin of the spin–spin splittings commonly measured in the nuclear magnetic resonance spectra of liquids.

The final examples we wish to use to illustrate the importance of these indirect couplings are ones which give rise to quadratic terms in the effective Hamiltonian; we can discuss them with the aid of Fig. III.11. The first is the second-order effect of the coupling between an applied field $\mathbf{B}$ and the orbital angular momentum $\mathbf{L}$, which mixes an excited electronic state with the ground state. The consequence of this second-order interaction is to produce a term in the effective Hamiltonian containing $\mathbf{B}^2$. This term is responsible for what is variously called "high-frequency paramagnetism", "temperature-independent paramagnetism" or "Van Vleck paramagnetism". It makes an important contribution to the magnetic susceptibility of all molecules, including closed-shell systems, since it does not depend upon the presence of electrons with unpaired spin.

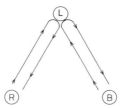

Fig. III.11. Second-order interactions leading to terms in the effective Hamiltonian which are quadratic in $\mathbf{R}$ or $\mathbf{B}$.

The second example is the second-order effect of the coupling between the rotational angular momentum $\mathbf{R}$ and the orbital angular momentum $\mathbf{L}$. Again this effect operates via excited electronic states, and in the effective Hamiltonian gives rise to a term containing $\mathbf{R}^2$. Remembering that the rotational angular momentum of the nuclei is also represented in the effective Hamiltonian by a term containing $\mathbf{R}^2$, we see the force of our opening remarks in this chapter concerning the rotational levels of CO. Although the rotational energy levels can be represented by the term $B\mathbf{R}^2 = BR(R + 1)$, the rotational constant B must be interpreted with more care than we exercised earlier, for it contains a direct contribution (certainly the major contribution) from the rotating nuclei, but also the indirect contribution discussed above arising from the electrons. If one is measuring very narrow lines to very high precision, an accurate interpretation must recognize the existence of both contributions.

Perhaps we have now said enough to be able to state what an effective Hamiltonian is; hopefully our earlier remarks in section III.2.1 will now be somewhat easier to appreciate. In summary the position is this. Leaving aside the relatively uninteresting translational motion, the eigenfunctions and eigenvalues of a molecule depend upon a number of nuclear and electronic properties. These are:

> electronic spatial energy,
> nuclear vibrational energy,
> nuclear rotational energy,
> electronic spin energy,
> nuclear spin energy,
> external field interactions.

If we want to determine "all" of the energy levels of the molecule, we must construct a Hamiltonian which includes explicitly all of these properties. If, however, we are studying the rotational levels of a molecule, we would find it excessively tedious and unrewarding to solve the complete problem every time. So instead we construct an effective Hamiltonian in which the rotational energy, electron and nuclear spin energy and external field interactions appear explicitly, but the effects of electronic spatial energy and vibrational energy are contained implicitly in the molecular constants appearing in the effective Hamiltonian. In rotational spectroscopy we are invariably dealing with a molecule in a particular vibronic state, that is, a particular vibrational level of a particular electronic state. Most frequently it is the ground vibronic state. So we can interpret our rotational spectrum in terms of an effective Hamiltonian for this vibronic state and obtain values for the molecular constants which appear in it. When we start to interpret the values of the constants, however, we must again turn to the complete Hamiltonian to see what those constants really represent.

Many different procedures for reducing the complete Hamiltonian to an appropriate effective Hamiltonian have been devised, but it is not necessary for us to go into the details, interesting though they are. It only remains to point out that the effective Hamiltonians employed in different branches of spectroscopy are usually quite different from each other. In the nuclear magnetic resonance of liquids, for example, the effective Hamiltonian will usually contain only the nuclear spin and external magnetic field explicitly, all other contributions to the molecular energy being reduced to constants, like the chemical shift and spin-spin coupling constants. In infrared spectroscopy, on the other hand, the effective vibrational Hamiltonian contains electronic spatial effects implicitly in the constants, and all other interactions should appear explicitly. The reason that effective vibrational Hamiltonians usually appear so simple is merely that the resolution is usually inadequate to reveal spin and external field effects, even if they are present.

The conclusion, therefore, is that one constructs an effective Hamiltonian appropriate to the experiment and with careful attention paid to the precision of the data. It is dangerous in the extreme to lift and use an effective Hamiltonian from the literature, without understanding fully the limitations which were imposed on its initial derivation.

III.3. Electronic States of Diatomic Molecules

The electronic states of diatomic molecules are constructed by considering the individual wave function for each electron and combining these wave functions in accordance with the Pauli exclusion principle. The spatial properties of a one-electron wave function in a diatomic molecule can be described in terms of three coordinates as shown in Fig. III.12. The z axis is the internuclear axis, ρ is the perpendicular distance of the electron from the z axis and ϕ is the azimuthal angle measured from a fixed plane through the internuclear axis (the zx plane). The potential energy depends only upon z and ρ and consequently the solutions to the Schrodinger wave equation can be written as a product of a function of z and ρ, and a function of ϕ alone:

$$\psi = \chi(z,\rho) . f(\phi). \tag{III.45}$$

The angular part of the wave equation is therefore separable and the solutions can be written in the form

$$\psi = \chi(z,\rho) \, e^{\pm i\lambda\phi} \tag{III.46}$$

where λ must be integral with values 0, 1, 2, 3, etc. The magnitude of the orbital angular momentum of the electron about the internuclear axis, k, is

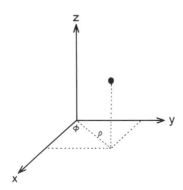

Fig. III.12. Electron spatial coordinates for a diatomic molecule.

equal to $\pm \lambda$ in units of $h/2\pi$ and we note from equation (III.46) that for $\lambda > 0$, there is always a two-fold degeneracy associated with the two opposite senses of the angular momentum. If $\lambda = 0$ the state (or orbital) is non-degenerate and the wave function is described as being σ-type. For $\lambda = 1$ the doubly degenerate wave function is π-type, for $\lambda = 2$ it is δ-type, and so on.

In a diatomic molecule with more than one electron, the total orbital angular momentum about the internuclear axis, Λ, is constructed by addition of the individual angular momenta, λ_i, for each electron i. We will see how this is done with the aid of some examples later. The nomenclature of the final electronic state is based on the resulting $|\Lambda|$ value according to the following table:

$$|\Lambda| = 0 \quad 1 \quad 2 \quad 3 \quad \text{etc.}$$

$$\text{electronic state} \quad \Sigma \quad \Pi \quad \Delta \quad \Phi \quad \text{etc.}$$

(III.47)

In addition to the orbital part of the wave function, we have also to consider the spin function. Each electron has a spin vector $\mathbf{s}_i$ equal to $\frac{1}{2}$ and two allowed spin orientations, $\pm\frac{1}{2}$, in units of $h/2\pi$. Vectorial addition of the individual spin vectors leads to a value of the total spin $\mathbf{S}$, and the total spin multiplicity, $2S + 1$, of the electronic state is written as a presuperscript. Thus for molecules in singlet spin states ($\mathbf{S} = 0$, $2S + 1 = 1$) we write $^1\Sigma$, $^1\Pi$, $^1\Delta$, etc. For molecules with $\mathbf{S} = \frac{1}{2}$ we have the doublet states, $^2\Sigma$, $^2\Pi$, $^2\Delta$, etc., for $\mathbf{S} = 1$ we have triplet states such as $^3\Sigma$, $^3\Pi$, and so on.

Two more symbols *can* appear in the nomenclature for total electronic states, and both relate to the spatial symmetry of the electronic wave function. Σ states will sometimes be written as Σ^+ or Σ^- states, the plus or minus signs denoting whether the total wave function is symmetric or antisymmetric with respect to reflection through a plane containing the internuclear axis (e.g. the xz plane in Fig. III.12). This distinction only arises for Σ states, because if Λ is greater than zero, the symmetry operation merely interconverts the two orbitally degenerate components. Moreover Σ^- states only occur if the outer electrons occupy orbitals with $\lambda > 0$, since a σ-type function is clearly invariant with respect to the reflection operation.

The final symbol which might appear is a post-subscript u or g, for example, a $^3\Sigma_g^+$ state. This symbol only appears for homonuclear diatomic molecules since it depends on whether the total wave function is symmetric (g) or antisymmetric (u) with respect to inversion in the centre of symmetry. In order to establish this, we need to know more about the wave function than we have discussed so far, and we will return to this point later.

Let us now consider a few examples which should make the application of these rules reasonably clear. Consider first nitric oxide, NO, which has a total of 15 electrons. Although the total orbital and spin angular momenta are constructed by consideration of all the electrons, in practice it is not necessary to concern ourselves with the closed electronic shells. In NO we are left with one outer electron which has a π-type wavefunction. Denoting the degenerate spatial functions $\pm\lambda$ by $(+1)$ and (-1), the four possible configurations, including electron spin, are

$$
\begin{array}{cc}
(+1) & (-1) \\
\\
\alpha & - \\
\\
- & \alpha \\
\\
\beta & - \\
\\
- & \beta
\end{array}
\qquad \text{(III.48)}
$$

where α and β denote the two possible spin states of the electron. Clearly $|\Lambda|$ is 1, and since $\mathbf{S} = \frac{1}{2}$ we have a $^2\Pi$ electronic state, which is doubly degenerate in both the orbital and spin parts of the wave function.

This is clearly a very simple case but oxygen with one more electron is more complicated. The possible electron configurations are as follows:

$$
\begin{array}{cccc}
(+1) & (-1) & |\Lambda| & \text{State} \\
\alpha_1 & \alpha_2 & 0 & \Sigma \\
\beta_1 & \beta_2 & 0 & \Sigma \\
\alpha_1 & \beta_2 & 0 & \Sigma \\
\beta_1 & \alpha_2 & 0 & \Sigma \\
\alpha_1\beta_2 & - & 2 & \Delta \\
- & \alpha_1\beta_2 & 2 & \Delta
\end{array}
\qquad \text{(III.49)}
$$

where the 1 and 2 denote the two outer electrons. The last two configurations satisfy the Pauli exclusion principle, in that the two electrons have the same spatial wavefunctions but different spins, and they constitute the two degenerate components of a $^1\Delta$ state. The other four configurations clearly belong to Σ states but in order to proceed further we must construct the appropriate wave functions in accord with the Pauli principle, which requires that the total wavefunction (space and spin) be antisymmetric when we permute the electrons 1 and 2. Now so far as the space part of the wave function is concerned we must allow each electron to have either spatial function, so that the two possible configurations are

$$(+1)_1(-1)_2 + (-1)_1(+1)_2$$
$$(+1)_1(-1)_2 - (-1)_1(+1)_2 \qquad \text{(III.50)}$$

The first function is symmetric with respect to interchange of the two electrons, whilst the second is antisymmetric. The spin functions for two electrons are

$$\left.\begin{array}{c} \alpha_1\alpha_2 \\ \beta_1\beta_2 \\ \alpha_1\beta_2 + \beta_1\alpha_2 \end{array}\right\} \quad S = 1 \qquad \text{(III.51)}$$
$$\alpha_1\beta_2 - \beta_1\alpha_2 \quad S = 0$$

The first three are clearly symmetric with respect to interchange of the electrons and they represent the three components of a triplet spin state with $S = 1$. The fourth function, which is antisymmetric, represents a singlet spin state with $S = 0$.

The complete wave function is now formed by combining each space function with a spin function such that it is overall antisymmetric. Hence the symmetric space function in (III.50) must be combined with the antisymmetric spin function in (III.51) so that the complete function may be written

$$(+1)_1^\alpha(-1)_2^\beta - (+1)_1^\beta(-1)_2^\alpha + (-1)_1^\alpha(+1)_2^\beta - (-1)_1^\beta(+1)_2^\alpha. \quad \text{(III.52)}$$

As we have already seen, this function with $S = 0$ and $|\Lambda| = 0$ represents a $^1\Sigma$ state. Reflection through a plane containing the internuclear axis interconverts the spatial functions $(+1)$ and (-1) so that under this operation (III.52) becomes

$$(-1)_1^\alpha(+1)_2^\beta - (-1)_1^\beta(+1)_2^\alpha + (+1)_1^\alpha(-1)_2^\beta - (+1)_1^\beta(-1)_2^\alpha. \quad \text{(III.53)}$$

This function is identical with (III.52), so that the electronic state must be $^1\Sigma^+$.

Consider now the triplet spin functions; any one of these can be combined with the antisymmetric space function in (III.50), yielding, for example,

$$(+1)_1^\alpha(-1)_2^\alpha - (-1)_1^\alpha(+1)_2^\alpha. \qquad \text{(III.54)}$$

Reflection through the symmetry plane converts this into

$$(-1)_1^\alpha(+1)_2^\alpha - (+1)_1^\alpha(-1)_2^\alpha \qquad \text{(III.55)}$$

which is equivalent to (III.54) except for an overall change of sign. Hence (III.54) represents a component of a $^3\Sigma^-$ state, and combination of the antisymmetric space function with the remaining triplet spin functions in (III.51) yields the other two components of the $^3\Sigma^-$ state.

We therefore see that the possible electron configurations shown in (III.49) give rise to three different electronic states, i.e. $^3\Sigma^-$, $^1\Sigma^+$ and $^1\Delta$. In molecular oxygen the $^3\Sigma^-$ state is the ground state, the $^1\Delta$ state is some 8000 cm^{-1} higher in energy, and the $^1\Sigma^+$ state is about 13,000 cm^{-1} higher. The difference in energy between these three states arises primarily from the electron–electron electrostatic interactions (V_{el} in equation III.18).

We have progressed to the point of describing the electronic states of diatomic molecules without invoking any particular model of the electronic structure, but to complete our discussion we should outline the molecular orbital theory which has proved to be of great value, particularly in discussing the molecular spectra. Molecular orbitals are constructed by combining atomic orbitals on the two atoms and Fig. III.13(a) shows the molecular orbital energy level diagram for homonuclear diatomic molecules containing first or second row elements. The bonding molecular orbitals themselves are illustrated in Fig. III.14. Starting from the lowest energy orbitals, the $\sigma_g(1s)$ and $\sigma_u(1s)$ molecular orbitals are formed by combining atomic $1s$ orbitals on each atom in-phase and out-of-phase respectively. The in-phase combination results in a bonding molecular orbital (i.e., it is of lower energy than the constituent atomic orbitals) and the subscript g denotes that it is symmetric with respect to inversion in the centre of symmetry. The out-of-phase combination is antibonding (i.e. higher in energy than the constituent atomic orbitals) and is of u symmetry (antisymmetric). These orbitals are clearly σ-type, that is, $\lambda = 0$. The bonding and antibonding $\sigma_g(2s)$ and $\sigma_u(2s)$ molecular orbitals are similar, except that they involve the $2s$ atomic orbitals.

The $2p$ atomic orbitals on the constituent atoms can combine with each other in two distinct ways. If the internuclear axis is labelled as the z direction, the $2p_z$ atomic orbitals can combine to form bonding and antibonding σ-type molecular orbitals which are clearly of g and u symmetry respectively. The $2p_x$ and $2p_y$ atomic orbitals then combine to form π-type molecular orbitals (i.e. $\lambda = 1$), the bonding orbital being antisymmetric (u) whilst the antibonding orbital is symmetric (g). It is not difficult to understand the degeneracy of these orbitals, since there is nothing which physically distinguishes the x and y directions.

The reader who studies Fig. III.14 carefully will observe that there are no symmetry reasons for excluding interaction between the $2p_z$ orbital on one atom and an s orbital (particularly the $2s$ orbital) on the other. Alternatively in terms of the molecular orbital diagram in Fig. III.13(a), the $\sigma_u(2s)$ and $\sigma_g(2p)$ molecular orbitals can interact with each other. The extent to which they do so depends upon the relative energies of the $2s$ and $2p$ atomic orbitals, but the result is to lower the energy of $\sigma_u(2s)$ and to increase the energy of $\sigma_g(2p)$. Clearly the possibility arises that the energy of $\sigma_g(2p)$ may increase sufficiently for it to lie above $\pi_u(2p)$, yielding the orbital diagram shown in Fig. III.13(b).

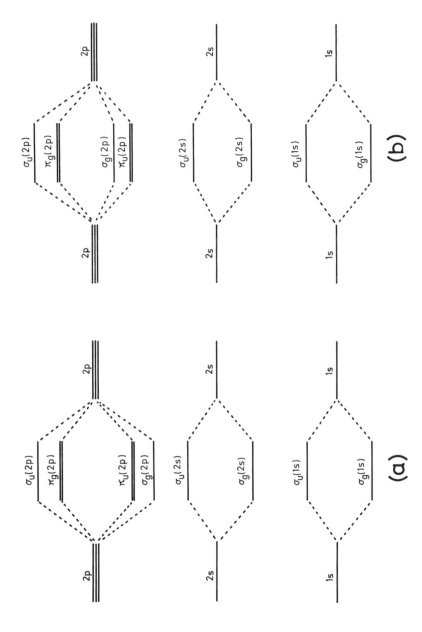

Fig. III.13. Molecular orbital energy level diagrams for homonuclear diatomic molecules.

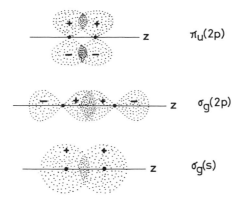

$\pi_u(2p)$

$\sigma_g(2p)$

$\sigma_g(s)$

Fig. III.14. "Shapes" of bonding molecular orbitals.

The question as to which orbital diagram applies to any particular molecule obviously becomes important when the highest energy electrons are to be placed in $\sigma_g(2p)$ or $\pi_u(2p)$.

The electron configurations of lowest energy for first and second row diatomic molecules are obtained by filling each molecular orbital in turn, in accord with the Pauli principle, until the available electrons are all accommodated. The configurations which have been established are as follows:

H_2 $\sigma_g(1s)^2$

Li_2 $\sigma_g(1s)^2\sigma_u(1s)^2\sigma_g(2s)^2$

B_2 . . . . . . $\sigma_u(2s)^2\pi_u(2p)^2$

C_2 . . . . $\pi_u(2p)^4$

N_2 . . . . $\sigma_g(2p)^2$

O_2 . . . . $\pi_g(2p)^2$

F_2 . . . . $\pi_g(2p)^4$

We see that, in fact, the diagram given in Fig. III.13(b) is found to be the correct one in every case where sufficient experimental information is available. For the molecule C_2, however, the $\pi_u(2p)$ and $\sigma_g(2p)$ orbitals are found to be very close in energy.

The energy level diagrams for heavier molecules are, of course, continuations of those shown in Fig. III.13, and they get progressively more complicated, particularly when d orbitals are involved. We will not pursue this development further, but will discuss individual molecules in Chapter IV. One further important point, however, is that we are now in a position to complete the nomenclature of the electronic states, that is, to indicate whether they are overall of g or u symmetry. Closed shells are overall symmetric (g), and for open-shell molecules we need consider only the outer electrons. Making use of the results $u \times u = g$, $g \times g = g$, $u \times g = u$ we see that the ground state of O_2, in which both outer electrons occupy g orbitals, is $^3\Sigma_g^-$; the excited states arising from the same electron configuration are $^1\Delta_g$ and $^1\Sigma_g^+$.

Heteronuclear diatomic molecules are less straightforward than the homonuclear because atomic orbitals of the same type on different atoms do not now have the same energy, and the molecular orbital energies depend upon the relative energies of the interacting atomic orbitals. Nevertheless the resulting diagram, particularly for atoms of similar atomic number, is not so different from those shown for homonuclear diatomic molecules. We can, in fact, use the molecular orbitals shown in Fig. III.13(b) to describe the electron configurations of a number of heteronuclear diatomic molecules. (Note that since a heteronuclear diatomic molecule has no centre of symmetry, the g and u classification is no longer relevant.) Some results are shown below, the asterisks denoting the antibonding orbitals.

$$\text{BN} \qquad \sigma(1s)^2\sigma^*(1s)^2\sigma(2s)^2\sigma^*(2s)^2\pi(2p)^3\sigma(2p)^1 \qquad : {}^3\Pi$$

$$\text{BO, CN} \quad . \quad . \quad . \quad . \quad . \quad . \quad . \quad \pi(2p)^4\sigma(2p)^1 \qquad : {}^2\Sigma$$

$$\text{CO} \qquad . \quad . \quad . \quad . \quad . \quad . \quad . \quad . \quad \sigma(2p)^2 \qquad : {}^1\Sigma$$

$$\text{NO} \qquad . \quad . \quad . \quad . \quad . \quad . \quad . \quad . \quad . \quad \pi^*(2p)^1 : {}^2\Pi$$

In concluding this rather brief account of the electronic structure of diatomic molecules, one final point is worth making. This point, which may well be in the reader's mind, is how these descriptions of the electronic wave function are related to the molecular Hamiltonian discussed in section III.2. A general answer to this question is that what we call the electronic wave functions are ideally eigenfunctions of that part of the Hamiltonian describing the kinetic energy of the electrons, the electron–nuclear and electron–electron electrostatic interactions, with the nuclear configuration fixed. Actually it is extremely difficult at the present time to calculate good wave functions for many-electron systems, although much current research is being directed at this problem. The molecular orbital description given in this section provides a satisfactory *language* in terms of which one may usefully discuss the electronic structure of diatomic molecules. However as a vehicle for quantitative description it does not take us very far. The molecular parameters which are determined by the high resolution experiments described in this book provide tests of molecular wave functions, but tests which are different in their sensitivity and detail. Some of the parameters determined seem at present to be impossible to calculate accurately from the best available wave functions, even for relatively simple species like CH and OH.

III.4. Coupling of Angular Momenta in Diatomic Molecules

Most readers will have some familiarity with the addition of vector forces, for we encounter examples every day. A body which is subject to two separate forces does not move in two different directions, but in a resultant direction whose magnitude and direction depends upon the magnitudes and directions of the individual forces. On the whole we are more familiar with problems of linear momentum, whereas the momenta in molecules which have occupied our attention in this chapter are mainly angular momenta. Nevertheless the problems to be solved are basically similar. A diatomic molecule whose constituent parts exhibit different angular momenta will itself possess a resultant total angular momentum. When more than two constituent angular momenta are present, however, there are several ways of describing their coupling to yield the total angular momentum. These different ways were first described by Hund, ánd are therefore known as the Hund's coupling cases. It should be emphasized at the outset that these coupling cases are idealized; any particular molecule will not necessarily conform exactly to any one of them, but it is usually closer to one than to any other. We shall see later that although the coupling cases can be represented by vector diagrams, their role can be made more quantitative.

First we review briefly the separate angular momenta to be considered, starting with the orbital angular momentum of the electrons. In an atom, which necessarily has spherical symmetry, the orbital angular momentum $\mathbf{L}$ is fully quantized. In a diatomic molecule, however, the axial symmetry allows the electrons to possess orbital angular momentum about the internuclear axis, but on the average the angular momentum about directions perpendicular to the axis is zero. Consequently only the component of $\mathbf{L}$ along the internuclear axis is a constant of the motion (neglecting spin effects for the moment) and this component is therefore well defined. This situation is summarized in Fig. III.15(a) and we recall that the different values of $|\Lambda|$ corre-

Fig. III.15. (a) Coupling of $\mathbf{L}$ to the internuclear axis. (b) Coupling of $\mathbf{L}$ and $\mathbf{S}$ to the internuclear axis.

spond to different electronic states which are usually widely separated in energy.

The next property to be considered is the spin of the electrons. We recall from the previous section that the spins of the individual electrons form a resultant S, and an associated magnetic moment μ_e. We also recall that the magnetic moments due to L and S interact (the spin-orbit coupling); since the component of L along the internuclear axis (Λ) is well defined, the magnetic moment due to L will lie along the internuclear axis and consequently the strong tendency of the spin-orbit coupling will be to orient the spin S so that the magnetic moment due to S also lies along the internuclear axis. In terms of the vector coupling diagram shown in Fig. III.15(b) we say that S also precesses about the internuclear axis, and its component along the axis is denoted by the symbol Σ. Hence we may also define the resultant component of total electronic angular momentum (orbital plus spin) along the axis as Ω, where Ω is given by

$$\Omega = \Lambda + \Sigma. \tag{III.56}$$

Now if $\Lambda \geq 1$ and $\Sigma \geq \frac{1}{2}$, Ω may have different values, which correspond to different so-called "fine structure" states. Suppose, for example, that $|\Lambda| = 1$ and $|\Sigma| = \frac{1}{2}$; the quantum numbers Ω, Λ and Σ can take the following values:

$$
\begin{aligned}
\Omega = \Lambda + \Sigma = \quad 1 + \tfrac{1}{2} &= \quad \tfrac{3}{2} \\
1 - \tfrac{1}{2} &= \quad \tfrac{1}{2} \\
-1 + \tfrac{1}{2} &= -\tfrac{1}{2} \\
-1 - \tfrac{1}{2} &= -\tfrac{3}{2}
\end{aligned}
$$

There are therefore two possible fine structure states, which we call $^2\Pi_{\frac{3}{2}}$ and $^2\Pi_{\frac{1}{2}}$, the post-subscript representing the value of $|\Omega|$. These two states are split apart by the spin-orbit coupling, the $^2\Pi_{\frac{3}{2}}$ state having the magnetic moments due to L and S parallel, the $^2\Pi_{\frac{1}{2}}$ state having them antiparallel. It is important to note that the spin-orbit coupling does *not* remove the two-fold Ω-degeneracy of the $^2\Pi_{\frac{3}{2}}$ and $^2\Pi_{\frac{1}{2}}$ states.

In molecules containing heavy atoms it is sometimes the case that the spin-orbit coupling is very much stronger than the coupling of L to the internuclear axis. In this case Λ is no longer a good quantum number and hence Σ is also not well defined. L and S are then coupled to form a resultant J_a which precesses about the internuclear axis with component Ω. If Λ is not defined, it follows that the electronic states cannot be labelled accurately as Σ, Π, Δ, etc. because the spin-orbit coupling mixes different states. This decoupling of L

from the internuclear axis by the strong spin-orbit coupling is illustrated in the vector diagram shown in Fig. III.16.

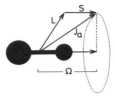

Fig. III.16. Coupling of **L** and **S** to form a resultant $\mathbf{J}_a$ which precesses about the internuclear axis.

So far we have discussed molecules in which $|\Lambda| > 0$ and we complete this discussion by now including the effects of the rotational angular momentum of the nuclei. The end-over-end rotation of the two nuclei in a diatomic molecule is represented by the vector **R** which is perpendicular to the internuclear axis. The coupling of **R** with **L** and **S** yields two of the Hund's coupling cases. Hund's case (a), shown in Fig. III.17 corresponds to coupling of **R** with **L** and **S** to yield the total angular momentum **J**, which is constant in magnitude and direction for given values of Ω and **R**. Hund's case (c), on the other hand, arises from coupling of **R** with $\mathbf{J}_a$. The essential difference between case (a) and case (c) is that in the former there is weak coupling between the three vectors **R**, **L** and **S**, whereas in the latter there is strong coupling between **L** and **S** to give a resultant $\mathbf{J}_a$ which is then weakly coupled with **R**.

If $\Lambda = 0$ (i.e. no orbital angular momentum) there is no reason why **S** should be coupled to the internuclear axis, which means that the projection Σ is not defined. This situation is described by Hund's coupling case (b), which may apply even when $\Lambda \neq 0$ if the spin-orbit coupling is very weak, as in very light molecules. Hund's case (b) is shown in Fig. III.18. The rotational angular momentum **R** is coupled to **L** to form a resultant **N**. **N** is then coupled to the spin vector **S** to form the total angular momentum **J**.

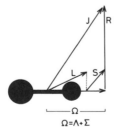

$$\Omega = \Lambda + \Sigma$$

Fig. III.17. Vector coupling diagram for Hund's case (a).

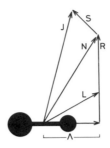

Fig. III.18. Vector coupling diagram for Hund's case (b).

There are two more coupling cases described by Hund, cases (d) and (e), but they are relatively unimportant and need not concern us here. The deeper significance of the Hund's coupling cases is, however, worth a little more thought. As we mentioned in section III.2, they correspond to particular coordinate transformations. Case (a), for example, corresponds to transformation of both the electron position and spin coordinates from a space-fixed axis system to the molecule-fixed rotating axes, whilst case (b) involves transformation of the electron position coordinates only. In the final section of this chapter (section III.6) we shall see that these coupling cases, apart from providing a meaningful language for discussion, form valuable starting points for the analysis and quantitative interpretation of molecular spectra.

In conclusion it is worth summarizing the angular momenta which we encounter in diatomic molecules, and defining again our notation, which conforms to international recommendation and practice:

L is the electronic orbital angular momentum
S is the electronic spin angular momentum
R is the rotational angular momentum of the nuclei
J is the total angular momentum of the molecule

$$\mathbf{J} = \mathbf{L} + \mathbf{S} + \mathbf{R}$$

N is the total angular momentum except for electron spin

$$\mathbf{N} = \mathbf{R} + \mathbf{L}$$
$$\mathbf{N} + \mathbf{S} = \mathbf{J}$$

Additionally, if nuclear spin angular momentum is present,

I is the nuclear spin angular momentum (for one nucleus)
F is the grand total angular momentum

$$\mathbf{J} + \mathbf{I} = \mathbf{F}$$

If both nuclei possess spin angular momentum, several coupling schemes are possible but we will leave these until we encounter actual examples later in the book.

III.5. Energy Levels Arising From Nuclear Rotation

III.5.1. Introduction

One of the themes of this chapter has been the manner in which the complex motions of the electrons and nuclei can be broken down into the sum of more readily recognizable dynamical properties, such as the vibrations and rotations of the nuclei. We have seen that complete separation of electronic and nuclear motion is not possible, and even without the electrons, it is not possible to separate uniquely the different sorts of nuclear motion. Nevertheless we saw that appropriate choices of origin and coordinate transformations take us some way and in this section we will consider in detail the rotation of the nuclei, which after all, constitute most of the mass of a molecule. At the same time we will note how the vibrational and rotational motions of the nuclei do interact, and how the electrons also have a role to play in the rotational properties of molecules.

Molecules come in all shapes and sizes, but among the properties which reflect the symmetry, size and mass of a molecule are its moments of inertia. Confining our attention to the nuclei, we know from classical mechanics that the moment of inertia of the nuclear masses about any axis we choose is given by

$$I = \sum_i m_i \rho_i^2 \tag{III.57}$$

where m_i is the mass of nucleus i and ρ_i is the perpendicular distance of nucleus i from the chosen axis. If we choose the origin of our coordinate system to lie at the centre of mass of the nuclei, we find, in general, that there is a unique choice for the direction of the three mutually perpendicular axes such that the moments of inertia about these axes are maxima or minima. These axes are called the "principal axes of the inertial tensor", and the corresponding moments of inertia about these axes are called the principal moments of inertia. Now we know that molecular shapes are not random but usually exhibit considerable spatial symmetry, and the major axes of symmetry in the molecule are always principal axes of the inertial tensor. Planes of symmetry in the molecule are always perpendicular to the principal inertial axes. Consequently the form of the inertial tensor is closely related to the geometry of the molecule, and this fact allows us to establish four main classes of molecule from the point of view of their inertial properties. They are listed below.

(i) *Molecules in which all three principal moments of inertia are different.* Such molecules are known as asymmetric tops or asymmetric rotors. An example is formaldehyde, where it is clear from Fig. III.19(a) that the moments of inertia about the three axes shown must be different. In this example only the z axis is actually a symmetry axis, but another of the principal axes must be perpendicular to the plane of the molecule, since the molecular plane is a symmetry reflection plane. The third principal axis is then necessarily perpendicular to the other two.

(ii) *Molecules in which two of the principal moments of inertia are equal, but different from the third.* Such molecules are known as symmetric tops, an example (Fig. III.19(b)) being CH_3Cl. The unique axis in this molecule is the three-fold axis along the C—Cl bond and obviously the axes perpendicular to the C—Cl bond direction must be equivalent, however we choose them. Symmetric tops almost always, in fact, contain an axis of three-fold or higher symmetry.

(iii) *Molecules in which the three moments of inertia are equal.* These are called spherical tops, an example being CH_4. As Fig. III.19(c) illustrates, the three four-fold axes are equivalent and are the principal axes of the inertial tensor. Spherical tops are not too common; they are relatively uninteresting to microwave spectroscopists, lacking an electric dipole moment because of their high symmetry.

(iv) *Molecules in which two moments of inertia are equal, and the third is zero.* Linear molecules fall into this class, the moment of inertia about the molecular axis being zero whilst, clearly, there is no unique choice of axes perpendicular to the molecular axis.

We will now discuss the rotational properties and energy levels of these four types of molecule, starting with the simplest which are, of course, the linear species.

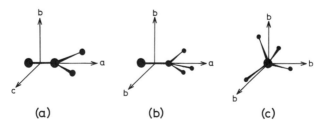

Fig. III.19. Directions of the principal inertial axes: (a) asymmetric rotor (e.g. H_2CO); (b) symmetric rotor (e.g. CH_3Cl); (c) spherical rotor (e.g. CH_4).

III.5.2. Linear molecules

We first confine our attention to diatomic molecules. As we saw in the previous sub-section, the entire rotational angular momentum of a diatomic molecule is due to the end-over-end rotation, the angular momentum vector thus lying perpendicular to the internuclear axis. The classical energy of rotation E_k about any axis k is given by

$$E_k = \tfrac{1}{2}I_k\omega_k^2 \tag{III.58}$$

where I_k is the moment of inertia about axis k and ω_k is the angular frequency of rotation. The classical angular momentum is simply $I_k\omega_k$, which we will call R_k, so that equation (III.58) may be written

$$E_k = \frac{R_k^2}{2I_k}. \tag{III.59}$$

Now in the case of the diatomic molecule rotation about the x and y axes is allowed, and since the moments of inertia about these axes are the same (see Fig. III.20), we obtain for the total classical rotation energy,

$$E = \frac{R_x^2}{2I_x} + \frac{R_y^2}{2I_y} = \frac{1}{2I}(R_x^2 + R_y^2) \tag{III.60}$$

where $I = I_x = I_y$ is called, simply, the moment of inertia. From equation (III.60) I is given by

$$I = m_1 r_1^2 + m_2 r_2^2 \tag{III.61}$$

$$= \frac{m_1 m_2}{m_1 + m_2} r^2 = \mu r^2 \tag{III.62}$$

where the origin is at the nuclear centre of mass, $r = r_1 + r_2$, and μ is called the reduced mass. Equation (III.60) becomes a quantum mechanical Hamiltonian describing the rotation if we regard R_x and R_y as operators and multiply them by $h/2\pi$. Hence

$$\mathscr{H}_{\text{rot}} = \frac{h^2}{8\pi^2 I}(R_x^2 + R_y^2). \tag{III.63}$$

Now since the square of the total rotational angular momentum $\mathbf{R}^2$ is equal to $R_x^2 + R_y^2 + R_z^2$, and R_z is zero for the linear molecule, equation (III.63) can be rewritten

$$\mathscr{H}_{\text{rot}} = \frac{h^2}{8\pi^2 I}\mathbf{R}^2 = \frac{h^2}{8\pi^2 I}R(R + 1) \tag{III.64}$$

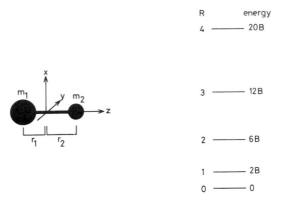

Fig. III.20. Nuclear rotation in a diatomic molecule and the lower rotational energy
levels.

in which we have replaced $\mathbf{R}^2$ by its eigenvalue, $R(R + 1)$. The quantity
$h^2/8\pi^2 I$ is called the rotational constant. Rotational constants are often
quoted by spectroscopists in units of cm^{-1}, in which case we divide by hc,
so that

$$B(cm^{-1}) = \frac{h}{8\pi^2 Ic}. \tag{III.65}$$

In equation (III.64) R is called the nuclear rotational quantum number and
takes all integral values 0, 1, 2, ... Hence the rotational energy levels occur at
$E = 0, 2B, 6B, 12B$, etc. as illustrated in Fig. III.20 and measurement of the
rotational level spacings yields a value for B. (Note that the symbol J often
appears in the rotational Hamiltonian instead of R. As we shall see below,
this is permissible if electronic orbital and spin effects are not present.)

The role of the electrons in molecular rotation is interesting but difficult to
appreciate. We saw in section III.4 that the total angular momentum of the
molecule, $\mathbf{J}$, may be expressed as the sum of the nuclear and electronic angular
momenta,

$$\mathbf{J} = \mathbf{R} + \mathbf{L} + \mathbf{S}. \tag{III.66}$$

Consequently the rotational Hamiltonian (III.64) may also be rewritten in the
form

$$\mathscr{H}_{\text{rot}} = B\mathbf{R}^2 = B\{\mathbf{J} - \mathbf{L} - \mathbf{S}\}^2 \tag{III.67}$$

and on expansion of the square in (III.67) we obtain terms like $2B\mathbf{J}.\mathbf{L}$ which
affect the *electronic* part of the total wave function. So we find that we cannot,
in fact, separate the rotational motion of the nuclei from electronic motions.

The effect of terms like $2B\mathbf{J}.\mathbf{L}$ is to change the rotational energy slightly by coupling different electronic states of the molecule. We shall have more to say about this effect, which is known as L-uncoupling, in the next chapter.

A further complication arises in that the previous discussion has assumed the molecule to be rigid, that is, that the nuclei remain the same distance apart. We know that this is not true because the nuclei vibrate; since the period of a vibration is much shorter than that of a rotation, however, we are able to accommodate the effects of vibration on the rotational Hamiltonian by using some mean value of the bond length r_v and hence rotational constant B_v for the vibrational state v considered, i.e.

$$\mathscr{H}_{\text{rot}} = B_v\mathbf{R}^2. \tag{III.68}$$

B_v is given by a series expansion in powers of $(v + \frac{1}{2})$, i.e.

$$B_v = B_e - \alpha_e(v + \tfrac{1}{2}) + \gamma_e(v + \tfrac{1}{2})^2 + \cdots. \tag{III.69}$$

The subscript e on the constants denotes their values for the equilibrium separation of the nuclei, r_e.

Having recognized that the molecule is not rigid, it is also not difficult to appreciate that as the molecule rotates, the nuclei experience a centrifugal force which tends to increase the distance between them, so that the apparent rotational constant decreases. This effect is accommodated by adding centrifugal distortion terms to the rotational Hamiltonian,

$$\mathscr{H}_{\text{rot}} = B_v R(R + 1) - D_v R^2(R + 1)^2 + H_v R^3(R + 1)^3 + \cdots, \tag{III.70}$$

the distortion constants D_v, H_v, etc. also being expressed as power series in the vibrational quantum number v, analogous to equation (III.69). Lest the impression be gained that these modifications to the expression for the rotational energy are made in an *ad hoc* manner, it should be said that careful studies of the interaction of vibration and rotation have provided sound justification for the procedures followed.

There are two other properties, both arising from symmetry considerations, which are important in the classification of rotational levels. The first is their so-called parity, which can be $(+)$-ve or $(-)$-ve according to the behaviour of the *total* eigenfunction on inversion of the space-fixed coordinates of all particles in the origin. A rotational level is of even $(+)$ or odd $(-)$ parity depending on whether the total eigenfunction remains unchanged or changes sign as a result of this inversion. Now the total wave function may be written as a product of electronic, vibrational and rotational wavefunctions,

$$\psi = \psi_{\text{el}}\psi_{\text{vib}}\psi_{\text{rot}} \tag{III.71}$$

and in order to determine how ψ behaves on inversion we look at the three product functions separately. The vibrational wavefunction is always unchanged on inversion since the inversion operation changes the *signs* of the coordinates but does not change the *magnitude* of the internuclear distance, on which ψ_{vib} depends. The electronic wavefunction ψ_{el} can, however, be of either $(+)$ or $(-)$ parity. The inversion operation is actually equivalent to the successive operations, (a), rotation through $180°$ about an axis perpendicular to the internuclear axis, followed by (b), reflection through a plane perpendicular to this rotational axis and containing the internuclear axis. The first of these does not affect ψ_{el}, but operation (b) is the same as that already described in section III.3. It defines Σ^+ and Σ^- states, but for Π, Δ, Φ, etc. states it merely interconverts the two orbital components. The rotational wavefunctions ψ_{rot} are multiplied by $(-1)^R$, or $(-1)^J$ if $S = 0$, on inversion. Consequently for a Σ^+ state the rotational levels are of even $(+)$ or odd $(-)$ parity depending on whether J is even or odd. For Σ^- states this rule is reversed, whilst for Π, Δ, Φ, etc. states each J level consists of two orbital components of opposite parity. If $S \neq 0$, the parities of the rotational levels for Σ states are usually governed by the value of N, the quantum number defined for Hund's case (b).

The determination of the relative parities of different rotational levels is important because uniform electric fields (oscillating or static) only connect levels of opposite parity; hence electric dipole transitions only occur between states of opposite parity.

The second important symmetry classification of the rotational levels of diatomic molecules arises for homonuclear molecules. We include the nuclear spin wavefunction in the total wavefunction, i.e.

$$\psi = \psi_{\text{el}}\psi_{\text{vib}}\psi_{\text{rot}}\psi_{\text{nuc}} \tag{III.72}$$

and note the Pauli exclusion principle which, stated more generally than before, demands the following behaviour on permutation of the two nuclei:

(i) the total wavefunction ψ must be antisymmetric if the nuclei behave as fermions (i.e. have spin $I = \frac{1}{2}, \frac{3}{2}, \frac{5}{2}$, etc.)

(ii) the total wavefunction ψ must be symmetric if the nuclei behave as bosons (i.e. have spin $I = 0, 1, 2, 3$, etc.).

Hence we must examine the behaviour of the total wavefunction (III.72) under the operation of permuting the positions of the two nuclei, P_{ab}, which essentially corresponds to relabelling them. We can achieve the same result as P_{ab} by a sequence of four operations (Fig. III.21) as follows:

1. Rotate the molecule through $180°$ about an axis perpendicular to the internuclear axis (this operation is called R_2).

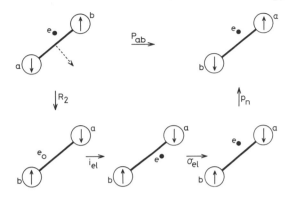

Fig. III.21. Permutation operation in a diatomic molecule. The nuclei are labelled *a* and *b* and an electron is labelled *e*. The electron position is shown as a closed circle if it is *above* the plane of the paper, and an open circle if it is *below*.

2. Invert the positions of the electrons in the origin (i_{el}).
3. Reflect the positions of the electrons through a plane perpendicular to the R_2 axis (σ_{el}).
4. Exchange the nuclear spins (p_n).

Consequently the permutation operator P_{ab} can be represented by the product of operations,

$$P_{ab} = p_n \sigma_{el} i_{el} R_2 \qquad\qquad (III.73)$$

and the operation of P_{ab} on the total wavefunction ψ can be represented as

$$P_{ab}\psi = (\sigma_{el} i_{el} \psi_{el})\psi_{vib}(R_2\psi_{rot})p_n\psi_{nuc}. \qquad (III.74)$$

Thus we have now to consider the sequence of operations written in (III.74).

(i) So far as the electronic wavefunction ψ_{el} is concerned, i_{el} multiplies it by $+1$ or -1 depending upon whether ψ_{el} is of *g* or *u* symmetry. We recall that we met this operation previously in section III.3. Likewise we have met σ_{el} earlier in this section; Σ^+ or Σ^- states are multiplied by $+1$ or -1 respectively, whilst the orbital components of Π, Δ, Φ, etc. states are interconverted.

(ii) ψ_{vib} is, as before, unaffected by any of the operations shown in Fig. III.21 because it depends only on the magnitude of the internuclear distance *r*.

(iii) R_2 changes the sign of ψ_{rot} by $(-1)^R$.

(iv) The operation of p_n on ψ_{nuc} is best illustrated by examples.

Suppose we are dealing with H_2, in which both protons have spin $I = \frac{1}{2}$. The possible nuclear wavefunctions are

$$a_\alpha b_\alpha$$

$$a_\beta b_\beta$$

$$\frac{1}{\sqrt{2}}(a_\alpha b_\beta + a_\beta b_\alpha)$$

$$\frac{1}{\sqrt{2}}(a_\alpha b_\beta - a_\beta b_\alpha)$$

p_n interchanges the labels a and b; the first three functions, constituting a spin triplet, are invariant, but the fourth changes sign.

We are now ready to handle the whole problem, and consider as an example H_2 in its $^1\Sigma_g^+$ ground state. Because the protons are spin $\frac{1}{2}$ particles, we require the result

$$P_{ab}\psi = -\psi. \tag{III.75}$$

Expanding (III.75) we find

$$P_{ab}\psi = (\sigma_{el}i_{el}{}^1\Sigma_g^+)\psi_{vib}(R_2\psi_{rot})p_n\psi_{nuc}$$
$$= ([+1][+1])^1\Sigma_g^+\psi_{vib}(-1)^R\psi_{rot}p_n\psi_{nuc}. \tag{III.76}$$

Hence in the lowest rotational level, $R = 0$, the nuclear wavefunction ψ_{nuc} must be the fourth (singlet) spin function if equation (III.75) is to be satisfied. In the second rotational level, $R = 1$, on the other hand, ψ_{nuc} must be a triplet spin function. In general, therefore, rotational levels with R even are associated with the singlet nuclear spin state (para-H_2), whilst rotational levels with R odd must be associated with the nuclear spin triplet state (ortho-H_2).

Now for a more complicated example, let us consider N_2 in its $^3\Sigma_u^+$ excited state; both nuclei are ^{14}N with spin $I = 1$, and consequently the total wavefunction must this time obey the requirement

$$P_{ab}\psi = \psi. \tag{III.77}$$

Expanding the total wavefunction as before we obtain

$$P_{ab}\psi = (\sigma_{el}i_{el}{}^3\Sigma_u^+)\psi_{vib}(R_2\psi_{rot})p_n\psi_{nuc}$$
$$= ([+1][-1])^3\Sigma_u^+\psi_{vib}(-1)^N\psi_{rot}p_n\psi_{nuc} \tag{III.78}$$

where N labels the rotational levels according to Hund's case (b). Since we are using case (b) functions, the electron spin function is invariant under the operations σ_{el} and i_{el}. Hence for rotational levels of even N, equation (III.77) is satisfied only if

$$p_n\psi_{nuc} = -\psi_{nuc}. \qquad (\text{III.79})$$

Conversely, for odd N it is necessary for the result

$$p_n\psi_{nuc} = +\psi_{nuc} \qquad (\text{III.80})$$

to be obeyed. We must therefore look at the nuclear spin states in more detail and classify them according to (III.79) or (III.80).

Each ^{14}N nucleus has spin $I = 1$ and hence three allowed spatial orientations corresponding to $M_I = +1, 0, -1$. Consequently there are nine possible basis functions representing the spins of both nuclei, which may be written,

$$a_1b_1, a_1b_0, a_1b_{-1}, a_0b_1, a_0b_0, a_0b_{-1}, a_{-1}b_1, a_{-1}b_0, a_{-1}b_{-1}. \qquad (\text{III.81})$$

In general these simple product functions do not have a definite permutation symmetry and we must therefore form appropriate linear combinations. The total nuclear spin $\mathbf{I}_T$ can be compounded by vectorial addition of the individual spins $\mathbf{I}_a$ and $\mathbf{I}_b$,

$$\mathbf{I}_T = \mathbf{I}_a + \mathbf{I}_b \qquad (\text{III.82})$$

which means that the possible values of I_T are $I_a + I_b$, $I_a + I_b - 1$, ..., $|I_a - I_b|$, that is $I_T = 2$, 1 or 0. It is not difficult to construct the correct linear combinations of the nine basis functions given in (III.81) and to classify them according to their I_T value. The results and permutation symmetries are:

$I_T = 2$ (symmetric) $M_I = 2 \qquad a_1b_1$

$$1 \qquad \frac{1}{\sqrt{2}}(a_1b_0 + a_0b_1)$$

$$0 \qquad \frac{1}{\sqrt{6}}(2a_0b_0 - a_1b_{-1} - a_{-1}b_1)$$

$$-1 \qquad \frac{1}{\sqrt{2}}(a_{-1}b_0 + a_0b_{-1})$$

$$-2 \qquad a_{-1}b_{-1}$$

$I_T = 1$ (antisymmetric)

$$M_I = 1 \qquad \frac{1}{\sqrt{2}}(a_1b_0 - a_0b_1)$$

$$0 \qquad \frac{1}{\sqrt{2}}(a_1b_{-1} - a_{-1}b_1)$$

$$-1 \qquad \frac{1}{\sqrt{2}}(a_{-1}b_0 - a_0b_{-1})$$

$$I_T = 0 \text{ (symmetric) } M_I = 0 \qquad \frac{1}{\sqrt{2}} (a_1 b_{-1} + a_{-1} b_1)$$

In the same way that we had a triplet and a singlet nuclear spin state in H_2, we now have a quintet ($I_T = 2$), a triplet ($I_T = 1$) and a singlet ($I_T = 0$) nuclear spin state for N_2. Clearly the $I_T = 2$ and 0 functions satisfy (III.80) and consequently are associated with those rotational levels for which N is odd. The $I_T = 1$ states satisfy (III.79) and are associated with rotational levels of even N.

The importance of these nuclear spin statistics is obvious. In very high resolution (e.g. molecular beam) studies of closed-shell molecules, or normal microwave studies of open-shell molecules, the different nuclear spin states have different energies because of nuclear hyperfine interactions, and consequently the form of the observed spectrum reflects the permutation symmetry. We have chosen to deal with $^3\Sigma_u^+$ N_2 because it has, indeed, been studied by molecular beam radio frequency spectroscopy, the details of which will be discussed in the next chapter.

The discussion in this section has been restricted to diatomic molecules so far, but much of it also applies to linear triatomic molecules. The rotational levels of a closed-shell linear triatomic molecule are given by equation (III.64), but the moment of inertia now depends upon three nuclear masses, m_1, m_2, m_3, and three internuclear distances r_{12}, r_{13}, r_{23} according to the formula

$$I = \frac{m_1 m_2 r_{12}^2 + m_1 m_3 r_{13}^2 + m_2 m_3 r_{23}^2}{m_1 + m_2 + m_3}. \qquad \text{(III.83)}$$

The parity arguments apply equally to the rotational levels of linear triatomics, and the permutation symmetry (and consequent nuclear spin statistics) also apply to molecules of the type X–X–X or Y–X–Y which possess a centre of symmetry. The important new considerations which arise for triatomic molecules concern their nuclear vibrations; a linear triatomic molecule has three different modes of vibration, compared with just one for the diatomic molecule. Vibrational-rotational and vibrational-electronic interactions can have profound effects upon the rotational levels of linear triatomic molecules, but we will reserve the details until the final chapter of this book, where we consider a few specific cases which have been studied.

III.5.3. Symmetric tops

A symmetric top molecule is one in which two of the principal moments of inertia are equal, but different from the third. Apart from rare cases of accidental equivalence, symmetric top molecules usually contain an axis of at least three-fold symmetry, this symmetry axis being the unique inertial axis, with the other two equivalent inertial axes being perpendicular to it. Two types of symmetric top are encountered, prolate and oblate. The two equal moments of inertia are conventionally called I_B or I^b. However, in the prolate top, of which CH_3Cl is an example, the moment of inertia about the unique symmetry axis is *smaller* than I^b and is conventionally labelled I_A or I^a. In the oblate top, an example being BCl_3 which is planar, the moment of inertia about the unique symmetry axis is *larger* than I^b, and in this case is labelled I_C or I^c (see Fig. III.22). The rotational constants are called A, B, C and are, of course, inversely proportional to the moments of inertia I^a, I^b and I^c respectively. The three principal inertial axes are labelled a, b and c.

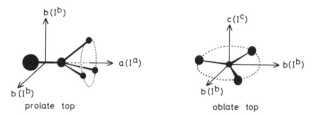

prolate top   oblate top

Fig. III.22. Examples of prolate (e.g. CH_3Cl) and oblate (e.g. BCl_3) symmetric tops.

The quantum mechanical Hamiltonian which describes the rotation of a prolate symmetric top molecule is

$$\mathscr{H}_{rot} = \frac{\hbar^2}{2I^b} \mathbf{J}^2 + \hbar^2 \left\{ \frac{1}{2I^a} - \frac{1}{2I^b} \right\} J_z^2 \qquad (\text{III.84})$$

and from this the prolate top rotational energies are found to be

$$E(J,K) = BJ(J + 1) + (A - B)K^2. \qquad (\text{III.85})$$

Alternatively, for an oblate top,

$$E(J,K) = BJ(J + 1) + (C - B)K^2 \qquad (\text{III.86})$$

K is a quantum number giving the value of $\mathbf{J}_z$, the angular momentum about the top axis, and for a given value of J, K takes integral values ranging from J to $-J$. We see from (III.85) and (III.86) that, except when $K = 0$, the rotational levels are doubly-degenerate. Figure III.23 illustrates the lower rotational levels for typical prolate and oblate symmetric tops; the main difference between the sets of levels is that, for a prolate top, $(A - B)$ in (III.85) is positive, whereas for an oblate top, $(C - B)$ in (III.86) is negative.

As in the case of diatomic molecules, the above expressions for the rotational levels have to be corrected to allow for centrifugal stretching, and for

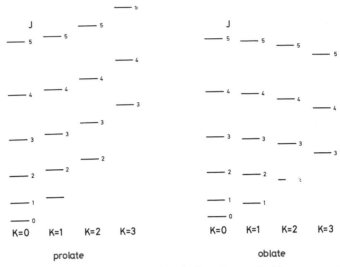

Fig. III.23. Lower rotational levels of typical prolate and oblate symmetric tops.

the dependence on the vibrational state of the molecule. A more accurate expression for the prolate symmetric top levels is

$$E_v(J,K) =$$
$$B_v J(J + 1) + (A_v - B_v)K^2 - D_J J^2(J + 1)^2 - D_{JK}J(J + 1)K^2 - D_K K^4$$
$$\text{(III.87)}$$

where the rotational constants B_v and A_v are given by

$$B_v = B_e - \sum_i \alpha_i^B(v_i + \tfrac{1}{2}d_i) + \cdots$$
$$A_v = A_e - \sum_i \alpha_i^A(v_i + \tfrac{1}{2}d_i) + \cdots.$$
$$\text{(III.88)}$$

The summations are taken over the normal vibrations of the molecule; d_i is the degeneracy of the normal mode i. Similar expressions hold for oblate tops. Vibration-rotation coupling is particularly important for molecules in degenerate vibrational states and actually leads to a splitting of the doubly-degenerate K levels. We shall not pursue this subject further, however, because there are as yet no examples in the free radical field.

Inversion symmetry raises a feature which is not encountered for diatomic molecules. For a non-planar symmetric top, inversion of the coordinates of all particles in the origin produces a configuration which is *not* alternatively obtainable by a rotation of the molecule. Consequently there are two forms of the molecule, which we may call left and right forms, which are usually separated by a high potential barrier so that the interconversion rate is slow. Since both forms have the same moments of inertia, their rotational levels are coincident and hence the levels shown in Fig. III.23 all possess a two-fold degeneracy arising from this so-called "inversion doubling". If the potential barrier between the two forms is low, however, interconversion can occur rapidly and the degenerate levels are then split; the wavefunctions of the new rotational levels correspond to equal mixtures of the left and right forms, and have either $(+)$-ve or $(-)$-ve inversion symmetry. The best known example of inversion doubling is NH_3; as yet there are no examples from the free radical field.

For a planar symmetric top, there is no inversion doubling because the inversion operation is now equivalent to a rotation of the molecule.

Nuclear permutation symmetry is also important for symmetric tops, and again leads to the important result that certain nuclear spin functions can only be associated with certain rotational levels. If and when symmetric top radicals are studied by high resolution spectroscopy, the observed nuclear hyperfine structure will depend critically on the nuclear spin statistics.

The final point to make is that for open-shell molecules with $S \neq 0$, each symmetric top rotational level will be further split by coupling between the rotational and electron spin moments. We shall learn more about this complication when we come to discuss asymmetric top molecules with unpaired electron spin.

III.5.4. Spherical tops

A spherical top molecule is one in which all three moments of inertia are equal; this is the case for molecules which have two or more axes of three-fold or higher-fold symmetry. The rotational levels are obtained from the symmetric top levels by putting $A = B$, so that their energies are given by the simple expression

$$E(J) = BJ(J + 1). \tag{III.89}$$

Again, corrections for non-rigidity and vibrational motions must be made along the lines described previously. Spherical top molecules are of relatively little interest to the microwave spectroscopist, however, because they lack an electric dipole moment, and although magnetic dipole transitions in an open-shell spherical top might be detectable, no such study exists at present so we shall say no more.

III.5.5. Asymmetric tops

An asymmetric top molecule is one in which all three principal moments of inertia are different; it follows that such molecules do not possess a three-fold or higher-fold axis of symmetry. The rotational Hamiltonian for an asymmetric top is

$$\mathscr{H}_{\text{rot}} = AJ_a^2 + BJ_b^2 + CJ_c^2 \tag{III.90}$$

in which $A > B > C$. The principal inertial axes are labelled a, b, c and Fig. III.19 shows an example, formaldehyde, with the orientation of these axes. The difficulty with (III.90) is that, unlike the symmetric top, it is not possible to obtain an exact closed-form expression for all the rotational energies of the asymmetric top. Two expressions for the energy levels have been found useful; the first, due to Wang (1929) gives the energies as

$$E(J_\tau) = \tfrac{1}{2}(B + C)J(J + 1) + [A - \tfrac{1}{2}(B + C)]W_\tau \tag{III.91}$$

and the second, due to Ray (1932) and to King, Hainer and Cross (1943) is

$$E(J_\tau) = \tfrac{1}{2}(A + C)J(J + 1) + \tfrac{1}{2}(A - C)E_\tau. \tag{III.92}$$

The parameters W_τ and E_τ depend upon a particular combination of A, B, C and J, and for a given value of J assume $2J + 1$ different values, labelled by the subscript τ. For slightly asymmetric tops, however, approximate closed-form expressions can be developed for the prolate and oblate symmetric top limits. For the near-prolate,

$$E(J,K) \simeq \tfrac{1}{2}(B + C)J(J + 1) + [A - \tfrac{1}{2}(B + C)]K^2 \tag{III.93}$$

whilst for the near-oblate top, we have

$$E(J,K) \simeq \tfrac{1}{2}(A + B)J(J + 1) + [C - \tfrac{1}{2}(A + B)]K^2. \tag{III.94}$$

These expressions are very useful for establishing the general pattern of the rotational levels, but they do not reveal the most interesting feature encountered in passing from the symmetric top to the slightly asymmetric top. This is that the K-degeneracy in the symmetric top is removed as the molecule becomes asymmetric. The splitting of the K-levels ($K \neq 0$) is called K-doubling or asymmetry-doubling. Figure III.24 shows how the rotational levels change as we pass from the prolate symmetric top on the left-hand side ($A > B = C$), through the asymmetric top range ($A > B > C$), to the oblate symmetric top limit ($A = B > C$) on the right-hand side of the diagram. We observe that levels of the same J value do not cross; however levels of different J can cross, although we have chosen not to illustrate this because the diagram then becomes rather more confusing. In the asymmetric top region the rotational levels can be labelled as J_{po} levels, where the subscript p refers to the K value in the prolate limit, and the subscript o refers to K in the oblate limit. In the alternative J_τ nomenclature, τ is obtained by subtracting the oblate K value from the prolate.

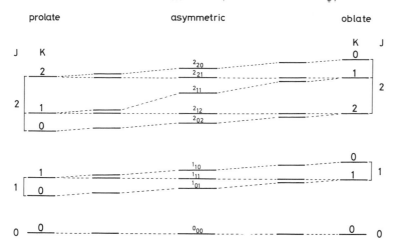

Fig. III.24. Correlation of the lower rotational levels of an asymmetric top with those of the prolate and oblate symmetric tops.

So far as the symmetry properties of the rotational levels are concerned, the possibility of inversion doubling arises for non-planar asymmetric tops, just as it does for the symmetric tops. However classification of the rotational levels according to the inversion behaviour of the total wavefunction is not commonly used. Instead it is usual to classify the rotational levels according to the behaviour of the rotational wavefunction with respect to two-fold rotations about each inertial axis. The details of this classification scheme are given by Herzberg (1945, 1966) and will not be repeated here.

Nuclear spin statistics are also important in many cases, but not in the asymmetric top open-shell molecules which have been studied so far. Finally if $S \neq 0$, the electron spin and rotational moments interact with each other, and for $S = \frac{1}{2}$, for example, each rotational level shown in Fig. III.24 is split into a doublet, the separation increasing with increasing J value. In this situation it is better to use the symbol N to label the rotational levels exclusive of spin, leaving J to label the final levels obtained after N and S have been coupled. Hence for $S = \frac{1}{2}$, each rotational level is split into a doublet with $J = N + \frac{1}{2}$ or $N - \frac{1}{2}$. We shall discuss the details in Chapter V when we deal with HCO and NO_2.

III.6. Calculation of Energy Levels

In concluding this chapter it is perhaps useful to look back to our remarks in section III.2.2 in the light of what has followed. In order to interpret a spectrum, we have to calculate the energies of the relevant levels and correlate the energy differences with transition frequencies. We recall that we can calculate the energy levels by setting up an appropriate effective Hamiltonian and then providing it with a suitable set of basis functions on which to operate. Let us briefly review the most important features of the effective Hamiltonian and the basis functions.

The effective Hamiltonian $\mathcal{H}_{\rm eff}$ contains terms describing the various angular momenta, the magnetic and electric interactions within the molecule, and interactions with external fields. In the analysis of microwave or radio-frequency spectra $\mathcal{H}_{\rm eff}$ is usually designed so that it operates within one vibrational level of one electronic state, and this is achieved by reducing the total "true" Hamiltonian to an $\mathcal{H}_{\rm eff}$ by some kind of perturbation method. The effects of terms which would otherwise couple different vibrational or electronic states are then contained in the "constants" appearing in $\mathcal{H}_{\rm eff}$. We must exercise judgement in the final choice of $\mathcal{H}_{\rm eff}$, including all the terms which are likely to have an observable effect on the spectrum; this, of course, involves some quantitative prior assessment of the various terms. If we have sufficient measurements we can determine the values of the "constants" appearing in $\mathcal{H}_{\rm eff}$, and then interpret those values by examining carefully the original reduction of the total Hamiltonian to $\mathcal{H}_{\rm eff}$. Many of these rather general remarks should be clearer after we have dealt with particular examples in Chapters IV and V.

Now what about the basis functions on which $\mathcal{H}_{\rm eff}$ is to operate? If $\mathcal{H}_{\rm eff}$ does not contain operators referring to vibrational and electronic motion, it is not necessary to specify vibrational or electronic quantum numbers in the basis functions. More usually the basis functions will specify the quantum numbers for molecular rotation, electron spin and nuclear spin, and we naturally choose the basis set which most nearly yields a diagonal matrix of $\mathcal{H}_{\rm eff}$. The matrix is actually infinite in size but can be truncated at a point depending upon the accuracy we desire in the final calculated energies.

For diatomic open-shell molecules we usually choose a basis set which corresponds to either Hund's case (a) or case (b). Hence with a case (a) basis set we specify the values of $J, \Omega, \Lambda, \Sigma, I$ and the spatial components M_J and M_I. With a case (b) basis set we specify J, N, S, I, M_J, M_I. For symmetric top or asymmetric top molecules we specify the values of J, K, S, N, I, and such space-fixed components as may be necessary. Each entry in the matrix of $\mathcal{H}_{\rm eff}$ (i.e. each matrix element) describes the effects of the various terms in $\mathcal{H}_{\rm eff}$ on the quantum numbers specified in the basis functions. The actual calculation of the matrix elements is fairly technical but straight forward. The final process of diagonalizing the matrix of $\mathcal{H}_{\rm eff}$ to obtain the energy levels is usually carried out by incorporating trial values for the constants, reducing the matrix to numerical form, and using a computer. The alternative process of obtaining explicit expressions for the energy levels in terms of the molecular constants is often difficult to achieve, particularly if $\mathscr{H}_{\text{eff}}$ contains more than a few terms. It may be that certain transition frequencies depend on only a restricted number of constants, in which case the spectrum can be analysed in stages. In general, however, the analysis of a complex, accurately measured spectrum, when many interactions are involved, is a complicated operation but a rewarding one.

IV. Studies of Diatomic Species

IV.I. Introduction

In this chapter we survey the results that have been obtained for diatomic species. They are arranged in what is nominally the increasing order of complexity, starting with $^1\Sigma$ states and concluding with $^3\Pi$ states. In practice the complexity in the analysis of a spectrum rather depends upon how close the electronic state is to one of the Hund's coupling cases. $^1\Sigma$ states, possessing neither spin nor orbital angular momentum, might be expected to be extremely simple and indeed they usually are if the $^1\Sigma$ state under consideration is the ground state. However, one of the examples we describe is H_2 in an excited $^1\Sigma$ state where, because of interactions with other nearby states, the molecule possesses a considerable magnetic moment and the spectral analysis is not straightforward.

IV.2. $^1\Sigma$ States

Our choice of $^1\Sigma$ states to be described must inevitably be somewhat subjective, but it is governed by the aim of illustrating as many spectral characteristics as possible before later plunging into all the additional complications which arise for molecules with large magnetic moments. We start with CS in its $^1\Sigma$ ground state, the experimental study of which was described in Chapter II. Kewley, Sastry, Winnewisser and Gordy (1963) have made measurements on the isotopic species $^{12}C^{32}S$, which is by far the most naturally abundant ($\sim 95\%$), and $^{12}C^{34}S$ ($\sim 4\%$ abundance). The ground electronic state of CS is a "good" $^1\Sigma$ state in every respect; it is well separated from, and therefore not significantly mixed with, any other electronic state. Electron spin and orbital contributions to the molecular magnetic moment are therefore negligible; in

148

addition nuclear spin effects are entirely absent, and the Hamiltonian determining the rotational levels contains only terms describing the nuclear rotation and centrifugal distortion. The appropriate effective Hamiltonian, given earlier in (III.70) is

$$\mathcal{H}_{rot} = E_{rot} = B_v J(J + 1) - D_v J^2(J + 1)^2 + H_v J^3(J + 1)^3. \quad \text{(IV.1)}$$

Kewley *et al.* measured the first five rotational transitions in $^{12}C^{32}S$, ranging from $J = 0 \rightarrow 1$ at 48·991 GHz to $J = 4 \rightarrow 5$ at 244·938 GHz. Using equation (IV.1) it is readily shown that the transition frequencies are given by

$$v = 2B_v(J + 1) - 4D_v(J + 1)^3 + H_v(J + 1)^3\{(J + 2)^3 - J^3\} \quad \text{(IV.2)}$$

and consequently there are three constants to be determined from five measurements. In the case of $^{12}C^{34}S$ the first three rotational transitions were studied.

We recall (equation (III.88)) that the constants in (IV.1) may be expressed as power series in $(v + \frac{1}{2})$, the constants in the expansion referring to the equilibrium configuration. In particular, equilibrium values B_e and D_e are obtained for CS, and these values may be related through the equation

$$D_e = 4B_e^3/\omega_e^2 \quad \text{(IV.3)}$$

where ω_e is the vibrational frequency. The value of ω_e obtained from this expression agrees well with the value obtained more directly from vibrational structure in the ultraviolet spectrum, showing that the analysis of the microwave data is internally self-consistent.

The microwave spectrum of $^1\Sigma$ CS is thus simple and readily interpreted. But because it is so simple, little or no information about the electronic structure of the molecule is obtained. The Stark perturbations produced by an electric field do, however, enable the electric dipole moment to be determined, as has been shown by Mockler and Bird (1955) who studied the $J = 0 \rightarrow 1$ transition. An applied electric field mixes adjacent rotational levels, which are of opposite parity, resulting in splitting of the $J = 1$ level and therefore of the microwave spectral line. The behaviour of the levels is illustrated in Fig. IV.1, in which we see a perfect example of a second-order Stark effect, the splitting being proportional to $(\mu E)^2$ and inversely proportional to B. Since the strength of the electric field (E) is known, the dipole moment is readily determined and found to be 2·0 Debye. It is interesting to note how large this is compared with CO, where the dipole moment is only 0·1 Debye. Recent experiments by McGurk, Tigelaar, Rock, Norris and Flygare (1973) have shown that the sign of the dipole moment in CS corresponds to the polarity C^-S^+.

The second example which we choose is AlF, studied by Lide (1963) using a high temperature Stark cell and generating the species by heating a mixture of Al and AlF_3 to 650 °C. The molecule has a $^1\Sigma$ ground state but the microwave rotational spectrum is more complicated than that of CS because of the presence of nuclear spin. In particular, the predominant isotope of aluminium is ^{27}Al, which has a spin I of $\frac{5}{2}$ and possesses a substantial electric quadrupole moment. The only naturally occurring isotope of fluorine (^{19}F) has spin $\frac{1}{2}$, therefore no quadrupole moment, and produces no observable hyperfine effects in the $^{27}Al^{19}F$ microwave spectrum.

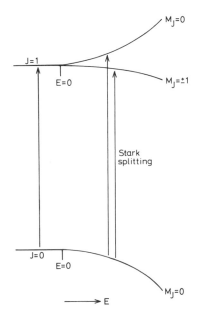

Fig. IV.1. Stark effect on the lowest rotational transition in CS.

The hyperfine components of the rotational levels can be characterized by values of what we shall call the "grand total angular momentum" quantum number F; $\mathbf{F}$ is formed by vectorial addition of $\mathbf{J}$ and $\mathbf{I}$ so that the possible values of the quantum number F are given by

$$F = J + I, J + I - 1, J + I - 2, \ldots, |J - I|. \qquad \text{(IV.4)}$$

Hence for the lowest rotational level of AlF, in which $J = 0$, $F = I$ and the only possible value of F is $\frac{5}{2}$. For the $J = 1$ level, however, the possible values of F are $\frac{7}{2}$, $\frac{5}{2}$ and $\frac{3}{2}$ and these three hyperfine components have different energies because of the electrostatic interaction between the nuclear quadrupole moment and the electric field gradient in which the nucleus is situated.

We discussed the physics of this interaction in section III.2.5, and it can be shown that the quadrupole energy of a level of given F is given by the expression

$$W_Q = -\frac{eqQ}{2I(2I-1)(2J-1)(2J+3)} \{\tfrac{3}{4}C(C+1) - I(I+1)J(J+1)\}$$

(IV.5)

where $C = F(F+1) - I(I+1) - J(J+1)$. Q is the nuclear quadrupole moment, q is the axial electric field gradient and the quantity eqQ is called the "quadrupole coupling constant". The energies of the three hyperfine components of the $J = 1$ level of AlF are found by substituting the appropriate values of F, J and I in equation IV.5. The results are:

$$W_Q(F = \tfrac{5}{2}) = 0 \qquad J = 0$$

$$\left.\begin{aligned}
W_Q(F = \tfrac{7}{2}) &= -\frac{1}{20}\,eqQ \\[1em]
W_Q(F = \tfrac{3}{2}) &= -\frac{7}{50}\,eqQ \\[1em]
W_Q(F = \tfrac{5}{2}) &= +\frac{4}{25}\,eqQ
\end{aligned}\right\} J = 1$$

(IV.6)

There are therefore three possible $\Delta J = \pm 1$ transitions, as shown in Fig. IV.2, and they all have electric dipole intensity. Clearly, the frequencies of these transitions yield values for eqQ and the rotational constant; the splittings between the lines depend upon the quadrupole coupling constant only, which was found to be -37.6 MHz. Quadrupole splitting is a characteristic feature

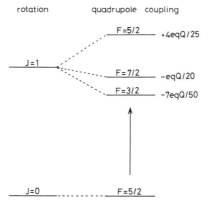

Fig. IV.2. ^{27}Al nuclear quadrupole splitting of the $J = 1$ rotational level in AlF.

of many microwave spectra and its analysis is straightforward in a well-behaved $^1\Sigma$ molecule.

The third example of a $^1\Sigma$ state which we discuss is from the work of Freund and Miller (1972) on an excited state of H_2. The principles and experimental details of their microwave-optical magnetic resonance studies were described in Chapter II, the excited state being the $G\ (3d\sigma)\ ^1\Sigma_g^+$ state of H_2 formed by electron bombardment of ground state H_2. Since a number of radiofrequency and microwave studies of electronically-excited hydrogen have been reported, we first outline briefly the nature of the electronic states of H_2.

The electronic states of H_2 could be described in terms of the molecular orbital diagram discussed in section III.3. However a somewhat different approach based on what is called the "united atom" model is in many ways more convenient. This model is particularly suitable for the excited states of H_2 where the internuclear distance is small compared with the orbital dimensions; the electronic states are constructed by considering the motion of a single outer electron in the electric field due to a core consisting of the two nuclei and the remaining electron. As the internuclear distance approaches zero, the eigenfunctions of the corresponding wave equation approach those of an atom and are characterized by values of two quantum numbers, n and l. These are entirely analogous to the principal and azimuthal quantum numbers of an atom; for a given value of n, l can take the values $0, 1, 2, \ldots, n-1$,

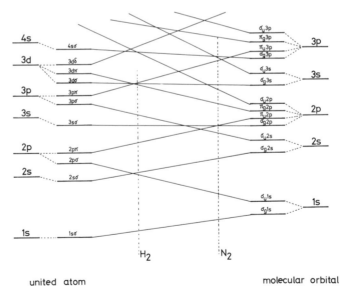

Fig. IV.3. Correlation of the energy levels derived from the united atom and molecular orbital models for a homonuclear diatomic molecule.

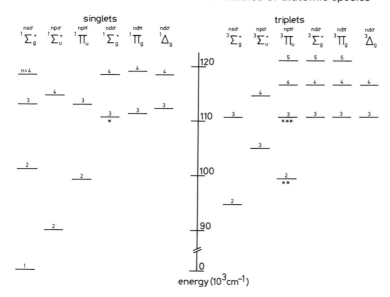

Fig. IV.4. The lower singly-excited singlet and triplet electronic states of H_2, based on the united atom model.

the corresponding functions being called s, p, d, ..., type. Each l level has, of course, a $2l + 1$ degeneracy which is partially split when the small separation of the nuclei is taken into account. The resulting levels are then additionally characterized by a quantum number $|\lambda|$ which takes values $0, 1, 2, \ldots, l - 1, l$, the levels being described as σ, π, δ, etc. Figure IV.3 shows a correlation diagram in which the energy levels based on the united-atom model are correlated with those based on the molecular orbital model. The total electronic states for H_2 are again constructed by placing each electron in an energy level, taking account of the Pauli principle.

We confine our attention to singly-excited states, that is, those in which one electron remains in the lowest energy level ($1s\sigma$) but the second is excited to a higher level. Figure IV.4 shows the lower excited singlet and triplet states which result, most of which have been identified by optical spectroscopy. Freund and Miller's work refers to the $(1s\sigma)(3d\sigma)$ G $^1\Sigma_g^+$ state; elsewhere in this chapter we shall describe molecular beam magnetic resonance studies of the $(1s\sigma)(2p\pi)$ c $^3\Pi_u$ state by Lichten (1960) and the $(1s\sigma)(3p\pi)$ d $^3\Pi_u$ state by Marechal and his co-workers (1972). There are still plenty of states left for the attention of others!

Freund and Miller's measurements refer to the $N = 1$ rotational level of the ground vibrational level of the G $^1\Sigma_g^+$ state. This level is of negative parity and is combined with a symmetric nuclear spin function. We recall that for two protons of spin $\frac{1}{2}$, we can form three symmetric combinations corresponding to total spin $I_T = 1$, and one antisymmetric combination with total spin $I_T = 0$. Unfortunately the line width obtained by Freund and Miller is too large to reveal proton hyperfine splittings.

Now the $G\ ^1\Sigma_g^+$ state of H_2 is not well-behaved in the sense that the ground state of CS was, the reason being that there are other nearby electronic states with which it can mix. The origin of the state mixing can be seen by considering the effective Hamiltonian used by Freund and Miller. If Hund's case (b) is used as a basis, the Hamiltonian is most appropriately written

$$\mathscr{H} = A_{\eta\Lambda} + B(\mathbf{N} - \mathbf{L})^2 - \mathscr{H}_{cd}$$
$$= A_{\eta\Lambda} + B\mathbf{N}^2 + B\mathbf{L}^2 - 2BL_z^2 - 2B(N_xL_x + N_yL_y) - \mathscr{H}_{cd} \qquad \text{(IV.7)}$$

where $A_{\eta\Lambda}$ represents the energy of the electronic state (exclusive of nuclear rotation); B is the nuclear contribution to the rotational constant, and $\mathscr{H}_{cd}$ represents the centrifugal distortion terms. The third and fourth terms affect only the overall energies of the electronic states but do not discriminate between different rotational levels. The second term, of course, represents the nuclear rotation and the fifth term represents coupling between the rotational and electronic motions. It is this term which is responsible for the mixing of different electronic states in the rotating molecule, and Freund and Miller show that the most important mixing is with nearby $^1\Pi_g$ and $^1\Delta_g$ states. Now these latter states have magnetic moments arising from electronic orbital angular momentum, and the rotational mixing with the $^1\Sigma_g^+$ state therefore confers a resulting magnetic moment to that state. Consequently an applied magnetic field readily removes the $2N + 1$ spatial degeneracy of the $N = 1$ level, and the resulting levels are characterized by values of M_N of $+1$, 0 and -1. Freund and Miller detected the two $\Delta M_N = \pm 1$ transitions shown in Fig. IV.5 by monitoring the fluorescence from the $G\ ^1\Sigma_g^+$ state to the $N = 0$ rotational level of a lower excited state, the $B\ ^1\Sigma_u^+$ state. Now the fluorescence radiation is polarized in a direction parallel to the applied magnetic field if M_N does not change in the fluorescence transition, but perpendicular if M_N does change. Consequently by monitoring the polarized fluorescence, microwave transitions between the M_N levels in the upper state can be detected because they change the relative populations of the three M_N levels and hence the polarized fluorescence intensity.

The effect of the applied magnetic field is represented in the Hamiltonian by the term $\mu_B\mathbf{B}.\mathbf{L}$; its main effect is to split the $2N + 1$ degeneracy, but it also has a small quadratic effect by mixing different N levels. The Zeeman effect is consequently not exactly linear and the two transitions should occur at slightly different field strengths. The spectrum obtained by Freund and Miller consists of one rather flat-topped line, which they analyse as two strongly overlapping components. At a microwave frequency of 9204·6 MHz, the line is observed at a field strength of 7383·0 gauss, which corresponds to an effective g value of 0·89077. The line width is determined by the radiative lifetime of the $G\ ^1\Sigma_g^+$ state, which is estimated to be 21 ± 4 nanoseconds. It is interesting to note that the g factor and the radiative lifetime are not unrelated, since both depend on the excited state wavefunction.

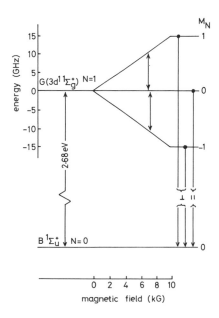

Fig. IV.5. Magnetic resonance and polarized fluorescence transitions in electronically-excited H_2 studied by Freund and Miller (1972).

Our final remark on this work is to comment that it is a splendid example of extracting the maximum information from a spectrum. Given one spectral line, the only measurable quantities are its position, shape and width. Each of these parameters yields information, as Freund and Miller have shown. It seems certain that their technique will prove to be a powerful way of learning more about excited electronic states, especially since the process of molecular excitation by electron bombardment is efficient, and suffers none of the selection rule restrictions which apply to normal optical (electric dipole) excitation.

IV.3. $^1\Pi$ States

The only current example of a high-resolution study of a $^1\Pi$ state is an excited $^1\Pi$ state of CS, studied by radiofrequency/optical double resonance. The experiments of Silvers, Bergeman and Klemperer (1970) and Field and Bergeman (1971) were described in Chapter II; their study is a fine example of the detailed information obtainable by this technique. The analysis is complicated, however, because of the close proximity of the $^1\Pi$ state to a number of other excited electronic states. Let us therefore first outline the rotational levels and expected properties of a $^1\Pi$ state which is well separated from any other electronic states.

We recall that a $^1\Pi$ state has $\Lambda = 1$ but $S = 0$, so that Σ is also zero. In terms of Hund's case (a), $\Omega = \Lambda = 1$ and in the lowest rotational level the value of the total angular momentum J is 1. The rotational Hamiltonian for the ground vibrational state may be written and expanded as follows:

$$
\begin{aligned}
\mathcal{H}_{\text{rot}} &= B\mathbf{R}^2 = B\{\mathbf{J} - \mathbf{L}\}^2 \\
&= B\{\mathbf{J}^2 - 2\mathbf{J}.\mathbf{L} + \mathbf{L}^2\} \\
&= B\{\mathbf{J}^2 - 2J_zL_z - 2J_xL_x - 2J_yL_y + L_z^2 + L_x^2 + L_y^2\} \\
&= B\{J(J+1) - \Lambda^2\} + B(L_x^2 + L_y^2) - 2B(J_xL_x + J_yL_y).
\end{aligned}
\tag{IV.8}
$$

The third line in the above expression is obtained by expanding $\mathbf{J}.\mathbf{L}$ and $\mathbf{L}^2$ in terms of components along the molecule-fixed axes x, y, z. The fourth line is obtained by replacing $\mathbf{J}^2$ by its eigenvalue $J(J+1)$ and noting that J_zL_z and L_z^2 may both be replaced by Λ^2 (because $J_z = L_z$, see equations (III.63) and (III.64)). Now the terms in Λ^2, L_x^2 and L_y^2 are independent of J and they do not affect the relative spacings of the rotational levels. In the absence of the last term in (IV.8), the rotational levels would be given simply by the term $BJ(J+1)$, and each would possess the two-fold Λ-degeneracy. The last term in (IV.8), however, represents coupling between the rotational and electronic motions, and it has the effect of mixing the $^1\Pi$ state with other electronic states of different Λ. This causes the Λ-degeneracy to be removed and the resulting splitting between components (the so-called Λ-doubling) increases with increasing J value. This situation is illustrated in Fig. IV.6. Provided the $^1\Pi$ state is well separated from other electronic states, the Λ-doubling is likely to be fairly small, particularly in the lower rotational levels.

Field and Bergeman (1971) have measured the Λ-doubling in the first nine rotational levels of the $A\,^1\Pi$ state of CS; it ranges from 13·6 MHz in the

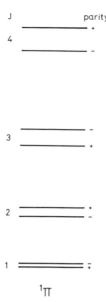

Fig. IV.6. Lower rotational levels of a $^1\Pi$ state, showing Λ-doublet splitting.

$J = 1$ level, to 1787·19 MHz in the $J = 9$ level. These measurements are made by inducing transitions between the Λ-doublet components of each J level in turn and monitoring the resulting changes in fluorescence from the $A\ ^1\Pi$ to the ground $X\ ^1\Sigma^+$ state. For the lowest rotational levels ($J = 1$ and 2), direct sweeps of the radiofrequency source were used, but for the higher rotational levels the spectra were recorded using a fixed radiofrequency and sweeping a d.c. electric field. The electric field changes the separation between the Λ-doublets, in a manner which we will discuss later. The first task, however, is to account for the size of the Λ-doubling.

In order to discuss the ground and excited states of CS, which possesses a total of 22 electrons, it is convenient to first consider the electronic configuration in terms of molecular orbital theory. The first 16 electrons occupy six σ-orbitals and one doubly-degenerate π-orbital, forming an inner core which need not concern us further. Of the remaining 6 electrons, two occupy a σ-orbital (which we call 7σ) and four occupy a π-orbital (which we call 2π) in the ground electronic state. The lowest vacant orbital in the ground state is another π-orbital, which we label as 3π. Hence the ground electronic configuration may be written

$$\text{(inner core)}\quad {}^{16}(7\sigma)^2(2\pi)^4(3\pi)^0 : X\ ^1\Sigma^+.$$

Excited states may be formed by $2\pi \to 3\pi$ or $7\sigma \to 3\pi$ excitations, or a combination of both. The $2\pi \to 3\pi$ excitation leads to an electron configuration $(7\sigma)^2(2\pi)^3(3\pi)^1$ and using the rules described in section III.3 it is not difficult to see that this configuration gives rise to Σ^+, Σ^- and Δ states of both triplet and singlet spin multiplicity. The $7\sigma \to 3\pi$ excitation, leading to the electron configuration $(7\sigma)^1(2\pi)^4(3\pi)^1$, gives rise to a $^3\Pi$ and a $^1\Pi$ state. It appears that the triplet states and the $^1\Pi$ state have similar energies and Fig. IV.7,

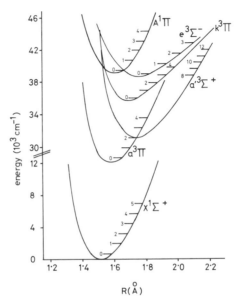

Fig. IV.7. Potential energy curves for the ground and lower excited electronic states of CS.

taken from Field and Bergeman's paper, shows approximate potential curves for the ground $^1\Sigma^+$ state and five excited states, including the A $^1\Pi$ state to which the measurements apply. It is clear from this diagram that the lowest vibrational level of the A $^1\Pi$ state ($v = 0$) lies close to excited vibrational levels of other electronic states, and that therefore strong interactions between the states are likely to occur. The ultraviolet spectrum of CS shows convincing evidence of perturbations between the states and in interpreting the radio-frequency measurements on the A $^1\Pi$ state, the effects of other electronic states must be taken into account. Field and Bergeman believe that the Λ-doubling in the A $^1\Pi$ ($v = 0$) state is, in fact, determined primarily by interaction with the a' $^3\Sigma^+$ ($v = 10$) state. The two states interact through spin-orbit coupling, but the Λ-doublet components of the $A\,^1\Pi$ state, which we label $A\,^1\Pi^{(+)}$ and $A\,^1\Pi^{(-)}$, experience slightly different interactions and therefore have different final energies; the difference is, of course, the Λ-doubling. In order to understand the details, it is necessary to discuss at some length the structure of $^3\Sigma$ states. This is deferred until section IV.7, where we will return to the question of spin-orbit mixing of $^3\Sigma$ with other states. Perhaps the most important general point to make, however, is that this is an example in which the analysis by means of an effective Hamiltonian operating entirely within a single vibronic state (the $A\,^1\Pi\,(v = 0)$ state) is not the best approach, the mixing of different vibronic states being all-important. This represents a breakdown of the Born-Oppenheimer approximation; such a breakdown is much more likely to occur for excited electronic states, which are often in close proximity.

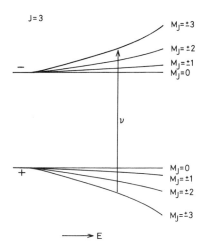

Fig. IV.8. Stark splitting of the Λ-doublet components of the $J = 3$ rotational level in $^1\Pi$ CS, and the transition studied by Field and Bergeman (1971).

Field and Bergeman recorded the Λ-doubling spectra of the $J = 3$ to 9 rotational levels by working at a fixed frequency and varying the magnitude of an applied electric field. The electric field mixes states of opposite parity and also partly resolves the $2J + 1$ spatial degeneracy of each level; the behaviour of the levels for $J = 3$ is shown in Fig. IV.8 and we see that the $\Delta M_J = 0$ transition between the $|M_J| = 3$ components can be tuned by sweeping the electric field. The transition frequency v is actually given by the expression

$$v = \{\delta^2 + [4\mu^2 E^2 M_J^2/h^2 J^2(J + 1)^2]\}^{\frac{1}{2}} \qquad (IV.9)$$

where δ is the Λ-doublet separation. Field and Bergeman measured the transitions for which $|M_J| = J$ and thus determined δ for the first nine J levels and μ for $J = 3$ to $J = 9$. Normally one would not expect the electric dipole moment to change noticeably from one rotational level to another, but in this case there is an increase from 0.6703 D $(J = 3)$ to 0.6786 D $(J = 9)$. Again it is the mixing of other electronic states which gives rise to this interesting variation.

This study illustrates most forcibly the fact that the neat classification of electronic states as $^1\Sigma$, $^1\Pi$, $^3\Sigma$, etc. is only an approximation. In many instances it works well, but we must always remember that it represents no more than a choice of basis set in which the total Hamiltonian is, hopefully, most nearly diagonal.

IV.4. $^1\Delta$ States

Four molecules in low-lying $^1\Delta$ states have now been studied, namely, O_2, SO, SeO and NF. We recall from the discussion of the electronic structure of O_2 in section III.3 that an electron configuration in which two electrons occupy degenerate π-orbitals gives rise to three electronic states, which are $^3\Sigma$, $^1\Sigma$ and $^1\Delta$. The wavefunctions for the $^1\Delta$ state were found to be

$$\frac{1}{\sqrt{2}} \pi_{+1}(1)\pi_{+1}(2)\{\alpha(1)\beta(2) - \beta(1)\alpha(2)\},$$

$$\frac{1}{\sqrt{2}} \pi_{-1}(1)\pi_{-1}(2)\{\alpha(1)\beta(2) - \beta(1)\alpha(2)\}.$$

These two functions are degenerate in the non-rotating molecule; in each case both electrons occupy the same orbital (i.e. have the same electronic spatial wavefunction) so that their spins must, by the Pauli exclusion principle, be paired. The degeneracy of the electronic wavefunctions is again called Λ-degeneracy. (Note that because the nuclear spin of ^{16}O is zero, for $^{16}O_2$ in its $^1\Delta$ state only *one* of the Λ-doublet components exists in each rotational level.)

The rotational structure of a $^1\Delta$ state is very simple, and its description follows closely our previous discussion of $^1\Pi$ states. Since $\Omega = \Lambda = 2$, the lowest rotational level has $J = 2$, successive higher levels have J values of 3, 4, 5, etc. The rotational Hamiltonian may be written

$$\mathscr{H}_{\text{rot}} = B\mathbf{R}^2 = B\{\mathbf{J} - \mathbf{L}\}^2 \tag{IV.10}$$

where we have made use of the fact that the total angular momentum $\mathbf{J}$ is simply equal to $\mathbf{R} + \mathbf{L}$. We expand equation (IV.10) as before and obtain

$$
\begin{aligned}
\mathscr{H}_{\text{rot}} &= B\{\mathbf{J} - \mathbf{L}\}^2 \\
&= B\{\mathbf{J}^2 - 2\mathbf{J}.\mathbf{L} + \mathbf{L}^2\} \\
&= B\{J(J + 1) - \Lambda^2\} + B(L_x^2 + L_y^2) - 2B(J_x L_x + J_y L_y).
\end{aligned}
$$
(IV.11)

The terms in Λ^2, L_x^2 and L_y^2 are independent of J and they do not affect the relative spacings of the rotational levels, which are given simply by the term $BJ(J + 1)$. The last term in (IV.11) represents rotational-electronic coupling and, as in the case of $^1\Pi$ states described previously, one of its effects is to remove the Λ-degeneracy of the states given in (IV.9). However, although this splitting is important in electronic Π states, it is very small in Δ states, particularly in the lower rotational levels. In all of the studies we shall describe, the Λ-doublet splitting is too small to be observed. In effect, therefore, the rotational levels are given very simply by the expression

$$
\mathscr{H}_{\text{rot}} = B_0 J(J + 1)
$$
(IV.12)

where $J = 2, 3, 4, 5$, etc. (see Fig. IV.9, which includes Λ-doublet splitting).
   Of the four species in $^1\Delta$ states so far studied, only $^1\Delta$ SO has been detected

Fig. IV.9. Lower rotational levels of a $^1\Delta$ state, showing Λ-doublet splitting.

by pure microwave rotational spectroscopy. $^1\Delta$ SO is generated by reacting carbonyl sulphide with the products of a microwave discharge in O_2 (Carrington, Levy and Miller, 1966a); it is known that discharged oxygen contains O atoms and $^1\Delta$ O_2 and it seems likely that $^3\Sigma$ SO is produced by the reaction

$$O + OCS \rightarrow CO + SO \, (^3\Sigma) \qquad\qquad (IV.13)$$

the $^3\Sigma$ SO then undergoing an exchange reaction with $^1\Delta$ O_2 (Carrington, Levy and Miller, 1966b; Breckenridge and Miller, 1972),

$$SO(^3\Sigma) + O_2(^1\Delta) \rightarrow SO(^1\Delta) + O_2(^3\Sigma). \qquad\qquad (IV.14)$$

Using this reaction with a parallel plate microwave spectrometer, Saito (1970) has detected the lowest rotational transition $J = 2$ to 3 at a frequency of 127,770·47 MHz. Using (IV.12) to give the rotational spacing, the B_0 value is determined to be 21,295·1 MHz, which agrees very well with an earlier value obtained from electron resonance work. In fact, equation (IV.12) should be corrected by the addition of a term representing the centrifugal distortion, $D_J J^2 (J + 1)^2$. There is at present insufficient data to provide a measurement of the distortion constant D_J but an estimated value (equation (IV.3)) of 30 kHz, which is probably fairly reliable, leads to a revised B_0 value of 21,294·6 MHz. From the B_0 value the bond length r_0 is calculated to be 1·493 Å. Saito also studied the Stark effect and determined the electric dipole moment; his value (1·31 Debye) again agreed well with that obtained from earlier electron resonance work, and we will discuss the Stark effect in more detail later.

The electron resonance spectrum of O_2 has been studied in the $J = 2$ and 3 rotational levels. SO has been studied in the $J = 2$, 3 and 4 levels, whilst SeO and NF in their lowest rotational levels have been investigated. In order to interpret the electron resonance results we must first consider the interaction between the molecule and an applied magnetic field; four different interactions, in decreasing order of importance, are:

(i) interaction of the magnetic field **B** with the magnetic moment $\boldsymbol{\mu}_L$ arising from the orbital motion of the electrons,
(ii) interaction of **B** with the magnetic moment created by the rotation of the molecule,
(iii) diamagnetic and higher-order paramagnetic interactions of **B** with the electrons (i.e. susceptibility effects),
(iv) interaction of **B** with any nuclear spin magnetic moments.

Of these interactions (i) is by far the most important and is represented in the effective Hamiltonian by the term

$$\mathscr{H}_{\text{Zeeman}} = g_L \mu_B \mathbf{B}.\mathbf{L}. \tag{IV.15}$$

Except for a very small relativistic correction, the orbital g_L factor has the value 1. The orbital Zeeman interaction has two main effects on the energy levels. First, each rotational level, characterized by its J value, is split into $2J + 1$ components, corresponding to the allowed $2J + 1$ orientations of the total angular momentum in the direction of the field $\mathbf{B}$. The energies of the components are, to first order, equal to $g_J M_J \mu_B B$, where the g value is given by

$$g_J = \frac{\Lambda^2}{J(J + 1)}. \tag{IV.16}$$

Hence for the first three rotational levels, $J = 2, 3, 4$, the g_J values are $\frac{2}{3}, \frac{1}{3}$ and $\frac{1}{5}$ respectively. The Zeeman splittings thus become smaller as J increases, as illustrated in Fig. IV.10(a).

The secondary effect of the orbital Zeeman term is to mix adjacent rotational levels, so that J is no longer a perfect quantum number. The extent of this mixing is inversely proportional to the spacing between the rotational levels being mixed, and proportional to the square of the magnetic field strength. Its effect is that the $2J + 1$ components of a given J level are not equally spaced as Fig. IV.10(a) suggests, but exhibit slightly unequal spacings for any particular field strength, as shown in Fig. IV.10(b). We will consider

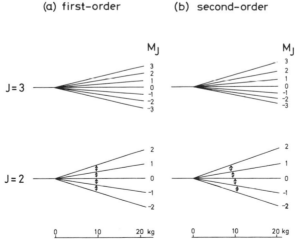

Fig. IV.10. Zeeman effect in the $J = 2$ and 3 rotational levels of a $^1\Delta$ state. (a) First-order. (b) Second-order.

the contribution to the Zeeman effect made by interactions (ii) and (iii) in due course, but first let us consider the form and content of the electron resonance spectrum if only the orbital Zeeman term is involved.

As we pointed out earlier, each J, M_J level shown in Fig. IV.10 is doubly-degenerate (Λ-degeneracy), and the two degenerate components are actually of opposite parity. An oscillating field applied perpendicular to the steady field B induces transitions between adjacent M_J levels provided it has the correct frequency. If the molecule possesses a permanent electric dipole moment, an oscillating electric field induces electric dipole transitions,

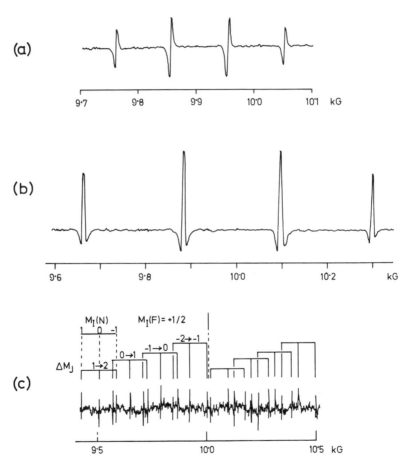

Fig. IV.11. X-band electron resonance spectra of (a) $^1\Delta\,O_2\,(J = 2)$, (b) $^1\Delta\,SO\,(J = 2)$, (c) $^1\Delta\,NF\,(J = 2)$.

$\Delta M_J = \pm 1$, between the components of opposite parity. Alternatively an oscillating magnetic field induces magnetic dipole transitions, which are between components of the same parity. Since electric dipole transitions are typically 10^2–10^4 times more intense than magnetic dipole transitions, the experiments on SO, SeO and NF were designed to permit the observation of electric dipole transitions. As we discussed in Chapter II, this means orienting the microwave cavity so that the microwave electric field is perpendicular to the steady field. However, in the case of $^1\Delta\, O_2$ which does not possess a permanent electric dipole moment, the electron resonance transitions are necessarily magnetic dipole and occur between states of the same parity. Since the Λ-doubling is negligibly small, the appearance of the electron resonance spectrum is the same, irrespective of whether the transitions are electric or magnetic dipole allowed; the intensities are so different, however, that with very few exceptions ($^1\Delta\, O_2$ being one), only electric dipole transitions have been detected in gas phase electron resonance spectra. This is in sharp contrast to electron resonance studies of condensed phases, where the transitions are invariably magnetic dipole allowed only.

In the lowest rotational level of a $^1\Delta$ state ($J = 2$), four transitions are possible, as shown in Fig. IV.10. If the second-order field effect were negligible, Fig. IV.10(a) shows that these four transitions would occur at the same field value, so that only one absorption line would be observed. At a typical X-band microwave frequency of 9 GHz, the g value of $\frac{2}{3}$ leads to a field value of ~ 9650 gauss for the transition. Now since the level spacings are not quite equal, due to the second-order mixing of rotational levels by the magnetic field, the four transitions are actually induced at separate field strengths and four lines are observed. These lines are expected to have relative intensities of $2:3:3:2$ and Figs. IV.11(a) and IV.11(b), showing the spectra of $^1\Delta\, O_2$ and $^1\Delta$ SO, illustrate that these expectations are realized. The experimental g values are 0·66663 for $^1\Delta\, O_2$ (Arrington, Falick and Myers, 1971) and 0·665 for $^1\Delta$ SO (Carrington, Levy and Miller, 1966a), very close to the first-order theoretical value of 0·6667. The separations between the four lines are inversely proportional to the rotational constant B_0, which is therefore readily determined. The electron resonance value of B_0 for $^1\Delta$ SO was, in fact, obtained prior to the microwave value given earlier.

The second rotational level ($J = 3$) is expected to yield an electron resonance spectrum of six lines, and the third rotational level ($J = 4$) should exhibit an eight line spectrum. The $J = 2$ and $J = 3$ spectra of $^1\Delta\, O_2$ have been measured (Arrington, Falick and Myers, 1971; Miller, 1971a) and the $J = 2, 3$ and 4 spectra of $^1\Delta$ SO have been measured (Uehara, 1971; Brown and Uehara, 1972). Since the first-order g values for $J = 3$ and 4 are $\frac{1}{3}$ and $\frac{1}{5}$ respectively, the spectra are observed at much higher magnetic fields if the microwave frequency is held constant, or alternatively at lower microwave frequencies for similar magnetic fields.

Let us now consider briefly the other contributions to the Zeeman effect. The interaction of the rotational magnetic moment with the applied field is usually written in the form

$$\mathscr{H} = -g_{\mathrm{r}}\mu_{\mathrm{B}}\mathbf{B}.(\mathbf{J} - \mathbf{L}) \qquad (IV.17)$$

where the rotational g factor g_{r} is composed of both nuclear and electronic contributions,

$$g_{\mathrm{r}} = g_{\mathrm{r}}^{\mathrm{n}} - g_{\mathrm{r}}^{\mathrm{e}}. \qquad (IV.18)$$

The main effect of this interaction is to shift the spectrum slightly, but not to change the separation between the lines. It is, in general, possible to use this shift to obtain the value of g_{r}, and since it is not difficult to calculate the nuclear contribution to the g factor, $g_{\mathrm{r}}^{\mathrm{n}}$, the measurement of g_{r} is usually taken to yield a value for the electronic contribution $g_{\mathrm{r}}^{\mathrm{e}}$. This electronic contribution arises from the fact that the electrons tend to follow the rotating nuclei, and therefore give rise to a magnetic moment similar in magnitude but opposite in sign to that arising from the rotation of the nuclei.

The diamagnetic interaction of the electrons and the applied field yields a term in the effective Hamiltonian which is quadratic in the field strength; we saw in Chapter III that there is also a paramagnetic contribution of similar form. The combined term, which has a small effect on the separation of the four lines, comprises a part which is isotropic (i.e. does not depend on molecular orientation) and a part which is anisotropic. It can be shown that the isotropic part does not affect the electron resonance spectrum but the anisotropic part does. Miller (1971a) has shown that by employing the value of $g_{\mathrm{r}}^{\mathrm{e}}$, it is possible to use the measured Zeeman effect to determine the anisotropy of the diamagnetic susceptibility, and from this the molecular quadrupole moment; this has been performed for $^1\Delta\,O_2$.

The interactions between $\mathbf{B}$ and nuclear spin magnetic moments can be quite substantial. However, they do not affect the electron resonance line positions to a significant extent and are therefore often omitted from the effective Hamiltonian.

We now come to consider nuclear hyperfine interactions. The dominant isotopes of oxygen and sulphur are ^{16}O and ^{32}S which have no nuclear spin. However the spectrum of $^1\Delta$ NF, shown in Fig. IV.11(c), is rich in hyperfine structure because ^{14}N has spin $I = 1$ and ^{19}F has $I = \frac{1}{2}$, both isotopes being present in almost 100% natural abundance. Interaction of ^{19}F results in a doublet splitting of each level shown in Fig. IV.10, corresponding to the two allowed orientations of the fluorine nucleus, which may be characterized by the quantum number $M_I^F = +\frac{1}{2}$ and $-\frac{1}{2}$. Interaction with the spin of ^{14}N leads to a further triplet splitting corresponding to the three allowed orientations of the nitrogen nucleus with $M_I = +1$, 0 or -1. Consequently the complete spectrum of $^1\Delta$ NF in the $J = 2$ rotational level arises from a quartet splitting due to the second-order Zeeman interaction, times a doublet splitting from hyperfine interaction with ^{19}F, times a triplet splitting due to ^{14}N. Happily, all expected 24 lines are resolved in the spectrum reported by Curran, MacDonald, Stone and Thrush (1973) and shown in Fig. IV.11(c); each line can be characterized by the M_J values involved and the quantum numbers M_I^N and M_I^F. The largest splitting is, in fact, the fluorine doublet splitting which separates the spectrum into two almost identical halves. The smallest splitting is the ^{14}N splitting.

Hyperfine splitting has also been observed in other $^1\Delta$ spectra. The spectrum of O_2 artificially enriched in ^{17}O has been studied (Arrington, Falick and Myers, 1971), and since ^{17}O has a spin of $\frac{5}{2}$, the hyperfine interaction results in a splitting into six hyperfine components. Similarly the spectrum of $^1\Delta$ SO can be obtained with a sufficiently good signal-to-noise ratio for hyperfine splitting from naturally abundant ^{33}S to be observed (Miller, 1971b). The abundance of ^{33}S is 0·8% and it has a spin of $\frac{3}{2}$. Consequently the spectrum of $^1\Delta$ $^{33}S^{16}O$ shows a quartet hyperfine splitting from the sulphur.

For those nuclei with spin greater than $\frac{1}{2}$, there are two possible mechanisms for hyperfine splitting in the $^1\Delta$ spectra, electric and magnetic. The electric quadrupole interaction arises in exactly the same way and produces the same kind of splitting in $^1\Delta$ states as we discussed earlier for $^1\Sigma$ states. However the quadrupole effects arising from ^{17}O and ^{33}S seem to be too small to be studied in $^1\Delta$ O_2 and SO, and the preliminary account of the $^1\Delta$ NF spectrum does not give a value for the ^{14}N quadrupole coupling constant.

Magnetic hyperfine interaction occurs via interaction of the orbital and nuclear spin magnetic moments. This hyperfine interaction can be written in the form

$$\mathscr{H}_{hf} = aL_z I_z \qquad (IV.19)$$

where the constant a depends essentially on the positions of the electrons responsible for the net orbital angular momentum. To be slightly more explicit, the magnitude of a depends upon $\langle 1/r^3 \rangle$ for the orbitally unpaired electrons, where r is the distance of the electron from the nucleus, and $\langle \; \rangle$ denotes the average over all possible positions of the electron. We see therefore that measurement of the magnetic hyperfine interaction provides rather direct information about the electronic wave-function; electric quadrupole constants provide similar information but it is usually rather less direct, because the coupling depends upon the total charge distribution.

If we add a homogeneous electric field to the magnetic field across the cavity, Stark splittings are produced in the spectra and from their analysis it is straightforward to determine electric dipole moments (Carrington, Levy and Miller, 1967b). Taking $^1\Delta$ SO as an example, consider the $M_J = 0 \rightarrow 1$ transition and the energy levels involved. As we have already discussed, each level possesses a two-fold Λ degeneracy, the degenerate components being of opposite parity. Consequently the single observed line arises from two super-imposed transitions between states of opposite parity $0(+) \rightarrow 1(-)$ and $0(-) \rightarrow 1(+)$. If a static electric field is now applied parallel to the magnetic field, it mixes the states of same M_J value but different parity, resolving the Λ-degeneracy. The two transitions now occur at different magnetic field values, leading to a Stark splitting in the spectrum. Provided the applied electric field is fairly small, two lines are observed and the separation between them depends on the strength of the electric field, the electric dipole moment, and the particular rotational level under investigation (see Fig. IV.12). Hence the dipole moment can be determined. Since the electric field mixes the states of opposite parity, transitions which initially have magnetic dipole intensity only (i.e. $(+) \rightarrow (+)$ and $(-) \rightarrow (-)$) gain electric dipole intensity as the electric field is increased and at sufficiently high field strengths are readily observable.

The Stark effects in $^1\Delta$ SO (Carrington, Levy and Miller, 1967b) and $^1\Delta$ SeO (Byfleet, Carrington and Russell, 1971) are essentially first-order because of the Λ-degeneracy and modest electric fields produce easily measurable Stark

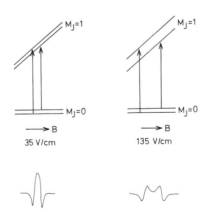

Fig. IV.12. Stark effect in the electron resonance spectrum of $^1\Delta$ SO ($J = 2$).

splitting, as shown in Fig. IV.12. The two Stark-split components are actually 180° out-of-phase because of the Stark modulation and detection technique employed. The magnitude of the electric field is usually estimated by measuring the Stark splitting in the electron resonance spectrum of NO, for which the dipole moment is known accurately.

In other cases where the Λ-doubling is resolved (for example, the $^2\Pi_{\frac{3}{2}}$ states of relatively light molecules) the Stark effect is second-order and the analysis slightly more complicated. However we will describe the details later in section IV.6.

IV.5. $^2\Sigma$ States

$^2\Sigma$ states currently occupy an intriguing position in microwave and radio-frequency spectroscopy. A conventional microwave or electron resonance experiment has yet to be described, but some very unconventional and fascinating studies have been reported. The CN radical, which has a $^2\Sigma^+$ ground state, has been detected in interstellar gas clouds through its microwave emission. An excited $^2\Sigma^+$ state of CN has also been studied by microwave/optical double resonance, the electronic excitation being achieved through chemical rather than optical pumping. An excited $^2\Sigma^+$ state of the OH radical (which has a $^2\Pi$ ground state) has also been investigated by microwave/optical double resonance. Finally, in a quite remarkable experiment, Jefferts has recorded a radiofrequency spectrum of the H_2^+ ion, the only high-resolution study of a molecular ion so far reported. H_2^+, of course, has a $^2\Sigma^+$ ground electronic state.

We first outline the main rotational features of a $^2\Sigma$ state, and then describe the experimental studies in the order given above. Most $^2\Sigma$ states conform closely to Hund's case (b) so that the nuclear rotational angular momentum $\mathbf{R}$ may be replaced by $\mathbf{N} - \mathbf{L}$ and the rotational kinetic energy is given by

$$\mathscr{H}_{\text{rot}} = B\mathbf{R}^2 = B(\mathbf{N} - \mathbf{L})^2. \tag{IV.20}$$

Since we are dealing with a Σ state, any effects due to $\mathbf{L}$ must be second-order and hence the rotational levels will be given by

$$\mathscr{H}_{\text{rot}} = B_0\mathbf{N}^2 = B_0N(N + 1). \tag{IV.21}$$

However, we must also consider the effects of the electron spin $S = \frac{1}{2}$. As we discussed in Chapter III, there is a magnetic coupling of the moments due to the rotation of the nuclei and the electron spin which is represented by the term

$$\mathscr{H}_{\text{sr}} = \gamma\mathbf{N}.\mathbf{S} \tag{IV.22}$$

where γ is called the spin-rotation coupling constant. In addition, however, we also showed how the $\mathbf{N} \cdot \mathbf{L}$ term arising from equation (IV.20) can combine with the spin-orbit coupling term $\mathbf{L} \cdot \mathbf{S}$ to give a term of the form $\mathbf{N} \cdot \mathbf{S}$ in the effective Hamiltonian. Consequently the constant γ contains a direct contribution from the rotating nuclei, and also an indirect second-order electronic contribution arising from admixture of excited electronic states with the ground state. In many instances the second-order contribution actually forms the major part of the constant γ.

The total angular momentum $\mathbf{J}$ is formed by vectorial addition of $\mathbf{N}$ and $\mathbf{S}$ so that the rotational levels may be characterized by quantum numbers N and J with $J = N + \frac{1}{2}$ or $N - \frac{1}{2}$. The effect of the spin-rotation term is to split the two spin components so that their energies are

$$J = N + \tfrac{1}{2}, \qquad E = B_0 N(N + 1) + \tfrac{1}{2}\gamma N,$$
$$J = N - \tfrac{1}{2}, \qquad E = B_0 N(N + 1) - \tfrac{1}{2}\gamma(N + 1). \qquad \text{(IV.23)}$$

This splitting is commonly referred to as spin-doubling (or formerly ρ-type doubling) and, as we see from equation (IV.23), it increases with increasing rotational quantum number N. Figure IV.13 shows the rotational structure of a typical $^2\Sigma^+$ state. All of the examples we shall discuss conform fairly well to this pattern but in much heavier molecules where spin-orbit coupling is

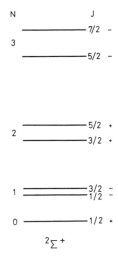

Fig. IV.13. Lower rotational levels of a $^2\Sigma^+$ state, showing the spin-doubling.

important and the rotational constants are small, case (c) coupling might become more appropriate.

The CN radical has been studied by ultraviolet spectroscopy (Herzberg, 1950) and has a $^2\Sigma^+$ ground state, with a rotational structure similar to that depicted in Fig. IV.13. The spin doubling constant γ has been determined to be 0.0070 cm^{-1} and the rotational constant B_0 is 1.8909 cm^{-1}. The rotational spectrum of CN in its ground state has not yet been detected in the laboratory but from the values of B_0 and γ obtained by analysis of the electronic spectrum, one predicts the following frequencies for the rotational transitions:

$$N = 0, \quad J = \tfrac{1}{2} \to N = 1, \quad J = \tfrac{1}{2} \quad \nu(\text{predicted}) = 113{,}165 \text{ MHz},$$

$$N = 0, \quad J = \tfrac{1}{2} \to N = 1, \quad J = \tfrac{3}{2} \quad \nu(\text{predicted}) = 113{,}480 \text{ MHz}.$$

$$(\text{IV.24})$$

Although laboratory spectroscopists have not succeeded in detecting the microwave spectrum, Penzias, Jefferts and Wilson (1972) have provided extremely good evidence for its identification through microwave emission from interstellar gas clouds in the Orion Nebula. The ^{14}N nucleus has spin $I = 1$ and hyperfine interactions with the electron spin, as well as an electric quadrupole interaction, are expected. Consequently the energy level diagram for the two lowest rotational levels is expected to be similar to that shown in Fig. IV.14, although it should be pointed out that since the hyperfine constants are not known, the magnitudes and signs of the splittings shown between the different F levels are only guesses. Penzias, Jefferts and Wilson

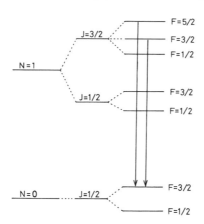

Fig. IV.14. Fine and hyperfine structure of the $N = 0$ and 1 rotational levels of CN in its $^2\Sigma^+$ ground state, and the transitions responsible for the observed interstellar emission lines.

detect two thermal (spontaneous) emission lines at frequencies (corrected for Doppler shifts) 113,492 MHz and 113,501 MHz which they attribute to the transitions indicated in Fig. IV.14. These frequencies are certainly close to the rotational frequency predicted in (IV.24) but until further lines are identified, one should perhaps keep an open mind on the interpretation of the interstellar spectrum. Ideally a laboratory spectrum is required to settle the matter, but this is easier said than done, and it may well prove easier in the future to study free radical spectra from interstellar space than to carry out successful experiments on earth!

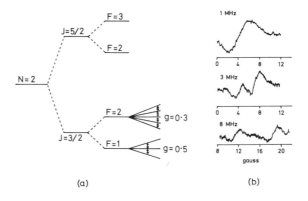

Fig. IV.15. Fine and hyperfine structure of the $N = 2$ rotational level in OH ($A\,^2\Sigma^+$), showing the double resonance transitions and spectra studied by German and Zare (1969).

We now turn to the terrestrial experiments of German and Zare (1969) who have studied the excited $A\,^2\Sigma^+$ state of the OH radical. The details of their experiment were described in Chapter II and here we concentrate on the energy levels involved. OH in its $^2\Pi$ ground state is excited to the $v = 0$, $N = 2$ rotational level of the $A\,^2\Sigma^+$ state, the excitation source being an emission line of zinc at 3072·06 Å. The appropriate energy level diagram for the $^2\Sigma^+$ state is shown in Fig. IV.15(a); there is a spin-doubling in the $N = 2$ level and the excitation process is actually to the $J = \frac{3}{2}$ spin component. Hyperfine interaction with the proton leads to a doublet splitting, the two levels being characterized, as shown, by F values of 1 and 2. German and Zare monitored the *polarized* fluorescence intensity from these levels and detected changes by removing the $2F + 1$ spatial degeneracy with an applied magnetic field and then inducing $\Delta M_F = \pm 1$ transitions with a radiofrequency field. We recall the similar experiments of Freund and Miller on excited H_2 where changes in the populations of different M_N levels induced by a microwave field resulted in changes of polarized fluorescence intensity. Figure IV.15(b) shows the double resonance spectra recorded by German and Zare for three different radiofrequencies and we see that at the higher frequencies, two lines are clearly resolved, corresponding to emission from the $F = 1$ and 2 levels. The main information to be gleaned from this spectrum is the g factors for the two levels. The experimental results and theoretical predictions for Hund's case (b) coupling are

$$F = 2, \qquad g_F(\text{exp}) = 0\cdot301 \pm 0\cdot0015, \qquad g_F(\text{theory}) = 0\cdot3,$$

$$F = 1, \qquad g_F(\text{exp}) = 0\cdot492 \pm 0\cdot0045, \qquad g_F(\text{theory}) = 0\cdot5.$$

Note that at the low field strengths used the Zeeman effect is linear so that different $\Delta M_F = \pm 1$ transitions within the same F level occur at the same field value.

Our third example again concerns the CN radical, but in its excited $B\,^2\Sigma^+$ state which has been studied through laboratory experiments. Again we described the experimental details in Chapter II and we recall that double resonance investigations are possible because of two particularly favourable features. The first is that reaction of N atoms with almost any organic compound produces CN in its excited $A\,^2\Pi$ state; the second is that the $v = 10$ level of this state is very close to the $v = 0$ level of the next excited state, the $B\,^2\Sigma^+$ state. Hence microwave transitions between the two states can be detected by monitoring fluorescence from the $B\,^2\Sigma^+$ state back to the ground state; the spectrum thus depends upon the rotational structure of both excited electronic states.

Figure IV.16 shows the energy level diagram appropriate to the experiments which have been performed. In the absence of any interaction between the $^2\Pi$ and $^2\Sigma^+$ excited states, we would expect each rotational level of the $^2\Sigma^+$ state to show spin-doubling because of the spin-rotation interaction, and each rotational level of the $^2\Pi$ state to show Λ-doubling, analogous to that described for the $^1\Pi$ and $^1\Delta$ states. Now in the case of CN, *one* spin-doublet component of the $^2\Sigma^+$ state interacts strongly with *one* Λ-doublet component of the $^2\Pi_{\frac{3}{2}}$ state, yielding two perturbed states which are labelled $\Sigma(p)$ and $\Pi(p)$ in Fig. IV.16. The other two states do not interact and therefore remain unperturbed; they are labelled $\Sigma(u)$ and $\Pi(u)$. The diagram refers to the $N = 4$ rotational level of the $^2\Sigma^+$ state in its ground vibrational level ($v = 0$) and the third rotational level of the $^2\Pi_{\frac{3}{2}}$ state in its tenth vibrational level ($v = 10$). The total angular momentum J has the value $\frac{7}{2}$ for the $^2\Pi_{\frac{3}{2}}$ rotational levels, and $\frac{7}{2}$ or $\frac{9}{2}$ for the spin components of the $N = 4$ level of the $^2\Sigma^+$ state. In addition there is a further triplet hyperfine splitting due to the spin $I = 1$ of ^{14}N, to give the final levels which are labelled by F and J.

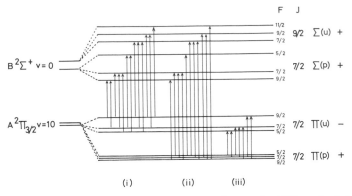

Fig. IV.16. Microwave transitions involving the $A\,^2\Pi_{\frac{3}{2}}$ and $B\,^2\Sigma^+$ excited electronic states of CN.

Microwave transitions are monitored by the resulting changes in fluorescence intensity from the $B\,^2\Sigma^+$ state to the $X\,^2\Sigma^+$ ground state. Clearly many transitions are possible but those which have been studied fall into three types as follows:

(i) $\Delta J = 0, +1$ transitions between $\Pi(u)$ and $\Sigma(u)$ or $\Sigma(p)$,
(ii) $\Delta J = 0, +1$ transitions between $\Pi(p)$ and $\Sigma(p)$ or $\Sigma(u)$,
(iii) $\Delta J = 0$ transitions between $\Pi(p)$ and $\Pi(u)$.

Transitions of type (i) were first studied by Evenson, Dunn and Broida (1964) and a spectrum is shown in Fig. IV.17. As we see from Fig. IV.16 the transitions also obey the selection rules $\Delta F = 0$ or ± 1 and although each individual transition frequency depends on properties of *both* electronic states, frequency differences between appropriate transitions yield information about the hyperfine structure of the $B\,^2\Sigma^+$ state alone. The transitions studied are all electric dipole allowed.

Transitions of type (ii) have been investigated more recently by Evenson (1969). They occur between states of the same parity, and are therefore magnetic dipole allowed only. Nevertheless their intensity is found to be comparable with that of the type (i) electric dipole transitions and one must conclude that the electric dipole matrix element between the Π and Σ states is very small. In this case the form of the spectrum is such that frequency differences between transitions yield direct information about the hyperfine interactions within the $^2\Pi$ state only.

Transitions of type (iii) have been studied by Evenson (1968) and by Pratt and Broida (1969). They are electric dipole transitions, confined to the Π state only. We shall meet many other examples of transitions between Λ-

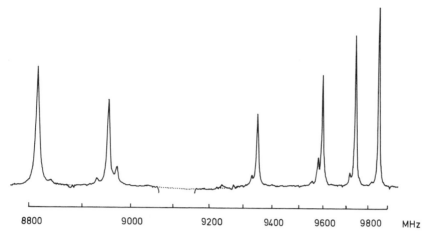

8800   9000   9200   9400   9600   9800   MHz

Fig. IV.17. Microwave/optical double resonance spectrum of CN, due to type (i) transitions (Evenson, Dunn and Broida, 1964).

doublet components in $^2\Pi$ states in the next section. In this example, the transitions occur in the frequency range 1570–1650 MHz.

Apart from knowledge of the lifetimes of the various states involved, the main structural information obtained from this work has been the determination of nuclear hyperfine constants in the $B\,^2\Sigma^+$ state. Radford (1964b) has analysed the data and determined the values of the constants which refer to the isotropic Fermi contact interaction and to the electron–nuclear dipolar interaction. He has also given a value for the ^{14}N electric quadrupole coupling constant (~ 5 MHz) and estimated the spin-doubling (separation between the $J = \frac{7}{2}$ and $\frac{9}{2}$ states) to be $\simeq 500$ MHz. The large number of papers describing studies of these electronic states in CN is testimony to the information obtainable, but also to the fact that this appears at present to be a rather unique system chemically. There are, of course, many examples from electronic spectroscopy of perturbations between excited electronic states, but rather few cases where the initial electronic excitation process is so efficient.

Finally we must describe some ingenious work by Jefferts (1969) on the H_2^+ ion. This is, of course, the simplest possible molecule, and Jefferts' work represents the only study by radiofrequency or microwave spectroscopy of a molecular ion. The principles behind the experiment are simple, however, and are as follows. H_2^+ ions are formed by electron impact on hydrogen at a pressure of 10^{-8} torr. The ions are held in a quadrupole r.f. trap where they are subjected to a small static magnetic field and at the same time irradiated with linearly polarized light applied parallel to the static field. The ions are excited from the ground electronic state to the excited $2p\sigma$ state, where they dissociate, i.e.,

$$H_2^+(1s\sigma) \xrightarrow{h\nu} H_2^+(2p\sigma) \longrightarrow H^+ + H. \qquad (IV.25)$$

The H^+ and remaining H_2^+ ions are removed from the trap and counted. Now the H_2^+/H^+ ratio depends upon the populations of the rotational and hyperfine levels of the ground state since the initial photoexcitation process follows normal optical selection rules. Consequently if the populations of these levels are changed in some way, the H_2^+/H^+ ratio changes. As in other double resonance experiments, the level populations can be changed by pumping transitions between the levels with electromagnetic radiation, the transitions in this case occurring in the frequency range 4–1300 MHz. The appropriate energy level diagram is shown in Fig. IV.18. The $N = 1$ rotational level is antisymmetric with respect to nuclear permutation and must therefore be combined with the symmetric proton spin function which, since each proton has spin $\frac{1}{2}$, corresponds to a total spin $I_T = 1$. On the other hand, the next rotational level $N = 2$ is symmetric and must therefore be combined with the antisymmetric nuclear spin function which has $I_T = 0$. States with $I_T = 1$ are called "ortho-H_2^+", those with $I_T = 0$ are "para-H_2^+". The para states are, of course, the simplest and for $N = 2$ we have only the usual spin-doubling caused by spin-rotation interaction. The resulting states may be labelled by quantum numbers $J \equiv F = \frac{3}{2}$ or $\frac{5}{2}$ (Fig. IV.18(*right*)). For the $N = 1$ ortho state, however, nuclear hyperfine effects are added to the electron spin doubling (because $I_T = 1$) and the states are then as shown in Fig. IV.18(*left*).

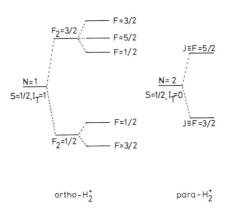

Fig. IV.18. Hyperfine structure of the $N = 1$ rotational level of ortho-H_2^+, and the $N = 2$ rotational level of para-H_2^+.

The quantum number F_2 is obtained by adding S and I_T,

$$\mathbf{F}_2 = \mathbf{S} + \mathbf{I}_T, \qquad F_2 = S + I_T, \ldots, |S - I_T| = \tfrac{1}{2} \text{ or } \tfrac{3}{2}$$

and F is obtained by adding F_2 and N, i.e.,

$$\mathbf{F} = \mathbf{F}_2 + \mathbf{N}, \qquad F = F_2 + N, , \ldots, |F_2 - N| = \tfrac{5}{2}, \tfrac{3}{2}, \tfrac{1}{2}.$$

In Jefferts' experiments the $2F + 1$ spatial degeneracy is removed by the static magnetic field (which is typically in the range 10–30 milligauss) and the transitions induced by the radio frequency field are of the type $\Delta F = \pm 1, 0$ and $\Delta M_F = \pm 1$. In the case of para-H_2^+ the relative energies of the $F = \tfrac{3}{2}$ and $\tfrac{5}{2}$ levels are determined only by the electron spin-rotation interaction, i.e. the effective Hamiltonian is just

$$\mathscr{H}_{\text{eff}} = \gamma \mathbf{S} . \mathbf{N} \qquad \text{(IV.26)}$$

and γ is determined by Jefferts to be 30·239 MHz for the $v = 5$ level and 28·092 MHz for the $v = 6$ level. The effective Hamiltonian for ortho-H_2^+ is more complicated because of nuclear spin terms and that used by Jefferts is

$$\mathscr{H}_{\text{eff}} = b\mathbf{I}.\mathbf{S} + cI_z S_z + \gamma \mathbf{S}.\mathbf{N} + f\mathbf{I}.\mathbf{N}. \qquad \text{(IV.27)}$$

For the $v = 4$ level the values of the constants are

$$b = 804 \cdot 065 \text{ MHz}, \qquad c = 98 \cdot 034, \qquad \gamma = 32 \cdot 636, \qquad f = 0 \cdot 038.$$

The effective Hamiltonians used are clearly zero-field Hamiltonians, the measurements being extrapolated back to zero field. The data are precise because the line widths are very small, typically 200–750 Hz.

In principle this technique can be applied to other simple ionic species but the requirements are fairly restrictive and it remains to be seen whether the method will be more generally applicable.

The final comment to be made about $^2\Sigma$ states of diatomic species is that, in many ways, they are excellent prototypes for the rotational structure and electron spin effects of non-linear radicals. The electron spin orientation is coupled to the molecular orientation only through the spin-rotation inter-action which is relatively weak. This means that electron resonance experiments are difficult because quite small magnetic fields decouple the electron spin from the molecular framework and it is then not possible to find electric dipole transitions whose frequencies can be tuned with an applied magnetic field. We will return to this crucial point in the next chapter.

IV.6. $^2\Pi$ States

IV.6.1. Introduction

Most of the diatomic radicals detected and studied by microwave techniques, particularly electron resonance, have been species in $^2\Pi$ ground states. There are two main reasons for this. Firstly, many diatomic radicals do have $^2\Pi$ ground states and secondly, one can usually predict fairly accurately where (in terms of magnetic field strength) the electron resonance spectra should occur.

The first step is to outline the main features of the rotational structure of $^2\Pi$ states. All of the species we shall describe are most conveniently discussed in terms of Hund's case (a). We recall from section III.4 that the essential features of case (a) coupling are (i), the orbital angular momentum $\mathbf{L}$ is strongly quantized with respect to the internuclear axis, the component along the axis being called Λ, and (ii), spin-orbit coupling is strong enough to result in quantization of the electron spin $\mathbf{S}$ with respect to the internuclear axis, the component along the axis being Σ. For a $^2\Pi$ state Λ can have values ± 1 and Σ has values $\pm \frac{1}{2}$. The component of total electronic angular momentum along the internuclear axis, Ω, can thus take the following values:

$$\Lambda + \Sigma \quad = \Omega$$
$$(+1) + (+\tfrac{1}{2}) = +\tfrac{3}{2}$$
$$(-1) + (-\tfrac{1}{2}) = -\tfrac{3}{2}$$
$$(+1) + (-\tfrac{1}{2}) = +\tfrac{1}{2} \qquad \text{(IV.28)}$$
$$(-1) + (+\tfrac{1}{2}) = -\tfrac{1}{2}$$

We see that $|\Omega|$ can be either $\frac{3}{2}$ or $\frac{1}{2}$ and the two corresponding states are called the "fine-structure states", $^2\Pi_{\frac{3}{2}}$ and $^2\Pi_{\frac{1}{2}}$. We also note the two-fold degeneracy of these states which is, of course, the Ω-degeneracy with which we are now familiar. In order to find the limitations of these pure case (a) descriptions, we must look at the Hamiltonian describing the spin-orbit coupling and the nuclear rotation. It has the form

$$\mathcal{H} = A\mathbf{L}.\mathbf{S} + B\mathbf{R}^2 \qquad \text{(IV.29)}$$

where B is, as usual, the rotational constant and A is an effective spin-orbit coupling constant. Actually this form of the spin-orbit interaction is only an approximate one, but it is good enough for most of the problems we shall deal with. We recall that the total angular momentum $\mathbf{J}$ is equal to $\mathbf{R} + \mathbf{L} + \mathbf{S}$ and consequently we may replace $\mathbf{R}$ in equation (IV.29) by $\mathbf{J} - \mathbf{L} - \mathbf{S}$ and expand the Hamiltonian in terms of components along molecule-fixed axes as follows:

$$\mathcal{H} = A\mathbf{L}.\mathbf{S} + B\{\mathbf{J} - \mathbf{L} - \mathbf{S}\}^2$$

$$= A\mathbf{L}.\mathbf{S} + B\{\mathbf{J}^2 + \mathbf{L}^2 + \mathbf{S}^2 - 2\mathbf{J}.\mathbf{L} - 2\mathbf{J}.\mathbf{S} + 2\mathbf{L}.\mathbf{S}\}$$

$$= A\Lambda\Sigma \tag{i}$$

$$+ B\{J(J+1) + S(S+1) - 2\Omega\Sigma - \Lambda^2\} \tag{ii}$$

$$- 2B(J_x S_x + J_y S_y) \tag{iii}$$

$$+ (A + 2B)\{L_x S_x + L_y S_y\} \tag{iv}$$

$$- 2B\{J_x L_x + J_y L_y\} \tag{v}$$

$$+ B(L_x^2 + L_y^2) \tag{vi}$$

$$\tag{IV.30}$$

The only term in (IV.30) which need not be considered seriously is term (vi) because it is effectively constant for a given electronic state and therefore does not affect the rotational level spacings.

Term (i), which represents the diagonal part of the spin-orbit coupling, is very important because it separates the two fine-structure states. For the $^2\Pi_{\frac{3}{2}}$ state the spin-orbit energy is, from (IV.28) equal to $+\frac{1}{2}A$ whilst for the $^2\Pi_{\frac{1}{2}}$ state the spin-orbit energy is $-\frac{1}{2}A$. Hence the fine-structure splitting is equal to A, which is therefore often called the fine-structure constant. Whether the $^2\Pi_{\frac{3}{2}}$ or the $^2\Pi_{\frac{1}{2}}$ state is the ground state depends upon the *sign* of A, and this in turn depends upon the electronic structure. As a general rule, a $^2\Pi$ state arising from an outer electron configuration $(\pi)^1$ has A positive, whilst for a $(\pi)^3$ configuration A is negative. (The rules are thus similar to those which apply to Russell-Saunders coupling in atoms.) Thus, for example, the CH radical which has seven electrons and an electron configuration $K(2s\sigma)^2(2p\sigma)^2 2p\pi$, where K represents the $1s$ electron inner shell, has a $^2\Pi_{\frac{1}{2}}$ ground state since A is positive. Such a state is called a "regular" $^2\Pi$ state. On the other hand the OH radical with an electron configuration $K(2s\sigma)^2(2p\sigma)^2(2p\pi)^3$ has a $^2\Pi_{\frac{3}{2}}$ ground state since A is negative; this is an example of a so-called "inverted" $^2\Pi$ state.

Term (ii) in (IV.30) gives the spacings of the rotational levels so that if only terms (i) and (ii) are important, we have the ideal case (a) molecule with a rotational structure similar to that depicted on the left-hand side of Fig. IV.19. Each rotational level still has the two-fold Λ-degeneracy indicated by equation (IV.28).

The remaining terms in equation (IV.30) have the effect of modifying this simple picture, and thus of making life more interesting. Term (iii) is called the "rotational distortion" term because it represents the tendency of the electron spin to become decoupled from the rapidly rotating nuclear framework. Its role as an operator in the effective Hamiltonian is to mix the $^2\Pi_{\frac{3}{2}}$

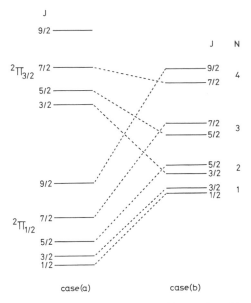

Fig. IV.19. Correlation of the case (a) and case (b) rotational levels of a $^2\Pi$ state.

and $^2\Pi_{\frac{1}{2}}$ states, and the extent to which it is successful depends upon the magnitudes of B and A. We recall that in Hund's case (b) coupling the electron spin is completely decoupled from the nuclear framework and the right-hand side of Fig. IV.19 shows the rotational levels for a $^2\Pi$ case (b) molecule, correlated with those for case (a). Light hydride molecules such as OH and CH, which have large rotational constants and relatively small A values, are intermediate between cases (a) and (b); heavier molecules like NO, ClO, etc. conform fairly closely to case (a) coupling. Even for OH and CH, however, it is more convenient to use case (a) as the basis for calculation and discussion. Note that each level still retains the two-fold Λ-degeneracy irrespective of whether the coupling is predominantly case (a) or case (b).

The two remaining terms in (IV.30) (terms (iv) and (v)) both contain the L_x and L_y operators and therefore they mix electronic states which differ in Λ values by $+1$ or -1. Consequently a $^2\Pi$ state is mixed with $^2\Sigma$ ($\Lambda = 0$) or $^2\Delta$ ($\Lambda = \pm 2$) states and the effect on the rotational levels of the $^2\Pi$ state is to remove the Λ-degeneracy, producing a splitting which is called Λ-doubling. As we shall see, Λ-doubling is a characteristic feature of light molecules, like CH and OH, which have large rotational constants. We shall examine the quantitative aspects of Λ-doubling in more detail later.

Apart from the terms given in equation (IV.30), the effective Hamiltonian usually contains terms describing nuclear hyperfine effects, and interactions with applied magnetic and electric fields. These will be discussed with respect to particular molecules later in this section.

IV.6.2. Molecules with large Λ-doubling

There are important qualitative and quantitative differences between the light molecules in which the Λ-doublings are quite large even for the lower rotational levels, and the heavier molecules in which the Λ-doubling is often unresolved. Although Λ-doubling makes the spectral analysis quite complex, we will deal with the light molecules first, starting with the OH radical which has been studied more extensively than any other radical species.

In the ground vibrational state the rotational constant B_0 for OH has the value $18\cdot52$ cm^{-1} and the fine-structure constant A is $-139\cdot7$ cm^{-1}. For the OD radical the rotational constant is somewhat smaller, $9\cdot87$ cm^{-1}. The lower rotational levels of OH and OD are illustrated in Fig. IV.20; we also

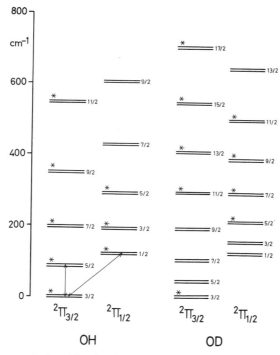

Fig. IV.20. Lower rotational levels of OH and OD. The asterisks denote those levels which have been investigated by high-resolution spectroscopy.

indicate the rotational levels which have been studied by high resolution spectroscopy. The transitions indicated have been detected by the laser electron resonance method described in Chapter II; the asterisks denote rotational levels which have been studied by pure microwave, electron resonance or beam maser experiments.

The rotational levels are, of course, further complicated by hyperfine interaction with ^{1}H in OH and ^{2}D in OD, with the result that each level in OH is split into a doublet whilst in OD each level is split into a triplet. In the case of OH each level may be characterized by the value of F which, for a given J level, is equal to $J + \frac{1}{2}$ or $J - \frac{1}{2}$. Leaving aside for the moment the rotational transitions between different J levels, all the microwave and electron resonance studies are concerned with transitions between Λ-doublets within a single J level, split by nuclear hyperfine interaction. In the pure microwave and beam maser experiments the transitions are studied by sweeping the microwave frequency. In Fig. IV.21(a) we show the transitions for the $^2\Pi_{\frac{3}{2}}$, $J = \frac{3}{2}$ level which can be investigated by means of swept-frequency techniques. Note that the transitions observed obey the selection rules $\Delta F = 0$ and $\Delta F = \pm 1$, although the $\Delta F = \pm 1$ transitions have a reduced electric dipole intensity. In Fig. IV.21(b) we show the same rotational level (without hyperfine splitting) as studied by electron resonance. The applied magnetic field removes the $2J + 1$ degeneracy and the resulting Zeeman components can be labelled by their M_J values; each level is further split into a doublet by the hyperfine interaction but this is omitted from the diagram to avoid complicating it too much. It is clear that the molecular constants which govern the energies of the levels in Fig. IV.21(a) are equally important in the electron resonance investigations, but in the latter the applied magnetic field provides an additional probe of the magnetic interactions. The zero-field constants are best obtained from the zero-field studies, and they can then be used to help interpret the electron resonance results and allow a more accurate determination of the Zeeman constants. It would be a tough proposition to determine all of the molecular constants appearing in the effective Hamiltonian from the electron resonance measurements alone.

The Λ-doubling frequencies (and hence microwave transition frequencies) range from approximately 1600 MHz in the lowest rotational level ($J = \frac{3}{2}$, $^2\Pi_{\frac{3}{2}}$) to 37,000 MHz in the $J = \frac{11}{2}$, $^2\Pi_{\frac{3}{2}}$ level (Dousmanis, Sanders and Townes, 1955). The most accurate measurements in the lowest rotational level have been made recently by Ter Meulen and Dymanus (1972) using the molecular beam maser method. The transition frequencies are

$$(F = 1)^- \leftrightarrow (F = 2)^+ \quad 1612.23101 \text{ MHz},$$

$$(F = 1)^- \leftrightarrow (F = 1)^+ \quad 1665{\cdot}40184,$$

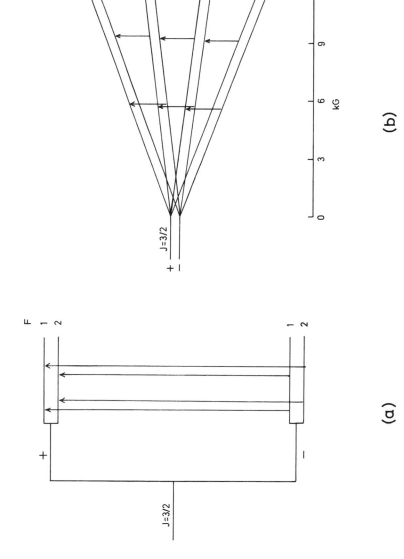

Fig. IV.21. (a) Pure microwave transitions in the $J = \frac{3}{2}$ ($^2\Pi_{\frac{3}{2}}$) level of OH. (b) Electron resonance transitions in the $J = \frac{3}{2}$ ($^2\Pi_{\frac{3}{2}}$) level of OH. (The proton hyperfine splitting is omitted.)

$$(F = 2)^- \leftrightarrow (F = 2)^+ \quad 1667 \cdot 35903,$$

$$(F = 2)^- \leftrightarrow (F = 1)^+ \quad 1720 \cdot 52998.$$

Such precision (~ 100 Hz) is possible because of the very narrow lines which molecular beam methods allow ($\sim 2 \cdot 5$ kHz widths), and knowledge of the frequencies to this level of accuracy is desirable because of astronomical studies. Many OH radio sources have been detected in interstellar space (and even in other galaxies) but the observed frequencies are always shifted somewhat from the rest transition frequencies because of the motion of the interstellar gas clouds and consequent Doppler shifts. Since determination of the velocities of the sources is of considerable importance to astronomers, accurate knowledge of the rest frequencies is indispensable.

Perhaps the best way of summarizing the results of the many studies of OH is to present the data in terms of the effective Hamiltonian, and to indicate the contributions made by different experimental studies to the determination of the molecular constants. First it is desirable to enlarge our earlier discussion of the Λ-doubling which is brought about by admixture of excited electronic states with the ground state through terms (iv) and (v) in equation (IV.30). As we mentioned earlier, these terms mix excited $^2\Sigma$ and $^2\Delta$ states with the $^2\Pi$ ground state and since the lowest excited state known for OH is the $A\ ^2\Sigma^+$ state described in the previous section, let us for the moment confine attention to the mixing of just that state with the ground state. In terms of case (a) descriptions, therefore, there are three electronic states ($^2\Pi_{\frac{3}{2}}$, $^2\Pi_{\frac{1}{2}}$, $^2\Sigma_{\frac{1}{2}}$) each with two-fold degeneracy, so that six states must be considered, namely,

$$\Pi_{\frac{3}{2}}, \ \Pi_{-\frac{3}{2}}, \ \Pi_{\frac{1}{2}}, \ \Pi_{-\frac{1}{2}}, \ \Sigma_{\frac{1}{2}}, \ \Sigma_{-\frac{1}{2}}.$$

Now it is very convenient to redefine the wave functions describing the states by introducing symmetric and antisymmetric combinations, so that

$$\Pi_{\frac{3}{2}}^{(+)} = \frac{1}{\sqrt{2}} \{\psi(\Lambda = 1, \Sigma = \tfrac{1}{2}, \Omega = \tfrac{3}{2}) + \psi(\Lambda = -1, \Sigma = -\tfrac{1}{2}, \Omega = -\tfrac{3}{2})\},$$

$$\Pi_{\frac{3}{2}}^{(-)} = \frac{1}{\sqrt{2}} \{\psi(\Lambda = 1, \Sigma = \tfrac{1}{2}, \Omega = \tfrac{3}{2}) - \psi(\Lambda = -1, \Sigma = -\tfrac{1}{2}, \Omega = -\tfrac{3}{2})\},$$

$$\Pi_{\frac{1}{2}}^{(+)} = \frac{1}{\sqrt{2}} \{\psi(\Lambda = 1, \Sigma = -\tfrac{1}{2}, \Omega = \tfrac{1}{2}) + \psi(\Lambda = -1, \Sigma = \tfrac{1}{2}, \Omega = -\tfrac{1}{2})\},$$

$$\Pi_{\frac{1}{2}}^{(-)} = \frac{1}{\sqrt{2}} \{\psi(\Lambda = 1, \Sigma = -\tfrac{1}{2}, \Omega = \tfrac{1}{2}) - \psi(\Lambda = -1, \Sigma = \tfrac{1}{2}, \Omega = -\tfrac{1}{2})\},$$

$$\Sigma_{\frac{1}{2}}^{(+)} = \frac{1}{\sqrt{2}} \{\psi(\Lambda = 0, \Sigma = \tfrac{1}{2}, \Omega = \tfrac{1}{2}) + \psi(\Lambda = 0, \Sigma = -\tfrac{1}{2}, \Omega = -\tfrac{1}{2})\},$$

$$\Sigma_{\frac{1}{2}}^{(-)} = \frac{1}{\sqrt{2}} \{ \psi(\Lambda = 0, \Sigma = \tfrac{1}{2}, \Omega = \tfrac{1}{2}) - \psi(\Lambda = 0, \Sigma = -\tfrac{1}{2}, \Omega = -\tfrac{1}{2}) \}.$$

$$(IV.31)$$

The reason for doing this is that we then need only to consider three functions at a time, and not all six. In other words, what would have been a 6×6 secular determinant factorizes into two 3×3 determinants because of the redefinition (transformation) of the basis wave functions so that they have a definite parity. The first of these 3×3 secular determinants involves the states $\Pi_{\frac{1}{2}}^{(+)}$, $\Pi_{\frac{3}{2}}^{(+)}$, $\Sigma_{\frac{1}{2}}^{(+)}$ and has the following general form:

$$
\begin{array}{c|ccc}
 & \Sigma_{\frac{1}{2}}^{(+)} & \Pi_{\frac{1}{2}}^{(+)} & \Pi_{\frac{3}{2}}^{(+)} \\
\hline
\Sigma_{\frac{1}{2}}^{(+)} & E_\Sigma - E_\Pi + F(J) - W & \mu & \eta \\
\Pi_{\frac{1}{2}}^{(+)} & \mu^* & -\tfrac{1}{2}A + F(J) - W & \varepsilon \\
\Pi_{\frac{3}{2}}^{(+)} & \eta^* & \varepsilon^* & +\tfrac{1}{2}A + F(J) - W
\end{array} = 0
$$

$$(IV.32)$$

On expansion of this determinant we obtain a cubic equation in the energy W (relative to the ground Π state) whose solutions give three possible values for W. The various quantities which appear in the secular determinant are as follows.

(i) $E_\Sigma - E_\Pi$ is the electronic excitation energy from the $^2\Pi$ ground state to the $A\ ^2\Sigma$ excited state (known from the electronic spectrum to be $32,682\cdot5$ cm^{-1}).

(ii) $F(J)$ simply indicates the rotational contributions to the energy which are, of course, functions of the value of J.

(iii) The matrix element μ is actually a sum of two matrix elements,

$$\mu = \theta + \zeta \qquad (IV.33)$$

where θ represents the coupling of Σ and Π states brought about by term (iv) in (IV.30) and ζ represents the coupling due to term (v), that is,

$$\theta = \langle \Pi | AL_x + 2BL_x | \Sigma \rangle$$
$$\zeta = f(J)\langle \Pi | BL_x | \Sigma \rangle \qquad (IV.34)$$

where $f(J)$ is a function of J.

(iv) The matrix element η also represents coupling due to term (v) in (IV.30) and may therefore be represented by

$$\eta = f'(J)\langle \Pi | BL_x | \Sigma \rangle \qquad (IV.35)$$

η and ζ are thus closely related.

(v) ε represents the rotational distortion term (iii) in (IV.30) which, as stated before, mixes the $^2\Pi_{\frac{3}{2}}$ and $^2\Pi_{\frac{1}{2}}$ fine structure states.

(vi) The asterisks simply denote complex conjugate elements which ensure that real eigenvalues are obtained.

The remaining three electronic functions $\Pi_{\frac{3}{2}}^{(-)}$, $\Pi_{\frac{1}{2}}^{(-)}$, $\Sigma_{\frac{1}{2}}^{(-)}$ are also mixed by the coupling terms in the effective Hamiltonian and the corresponding 3×3 secular determinant is identical with (IV.32) except that the element μ is given by

$$\mu = \theta - \zeta. \tag{IV.36}$$

Now suppose we want to derive an expression for the Λ-doubling in a particular J level of the $^2\Pi_{\frac{3}{2}}$ state. We first solve the determinant (IV.32), putting in the appropriate J values and obtaining three values of W, one of which, $W_{\frac{3}{2}}^{(+)}$, corresponds to the "final" energy of the state we still call (despite the mixing) $\Pi_{\frac{3}{2}}^{(+)}$. We follow the same procedure for the second determinant involving $\Pi_{\frac{3}{2}}^{(-)}$, $\Pi_{\frac{1}{2}}^{(-)}$, $\Sigma_{\frac{1}{2}}^{(-)}$ and obtain the value of $W_{\frac{3}{2}}^{(-)}$ corresponding to the state $\Pi_{\frac{3}{2}}^{(-)}$. The Λ-doubling in the state concerned is then $W_{\frac{3}{2}}^{(+)} - W_{\frac{3}{2}}^{(-)}$ (or $W_{\frac{3}{2}}^{(-)} - W_{\frac{3}{2}}^{(+)}$). It is, of course, clear that the expression for the Λ-doubling frequency will contain all the quantities which appear in (IV.32) and, as is the way with cubic equations, will be an unedifying mess. Also, although the theory as constructed above mentions only one $^2\Sigma$ excited state, the measured Λ-doubling frequency depends on all possible Σ states which can be mixed with the ground state. However the essential features can be picked out and one finds that the Λ-doubling frequency for any given J level depends upon the values of six main quantities. They are:

(i) the electronic excitation energy (or energies), $E_\Sigma - E_\Pi$,

(ii) B_Π, the rotational constant in the $^2\Pi$ state ($18 \cdot 515 \, \text{cm}^{-1}$ for OH in its ground vibrational level),

(iii) B_Σ, the rotational constant in the $^2\Sigma$ state ($16 \cdot 961 \, \text{cm}^{-1}$ for OH),

(iv) the rotational distortion element ε which is proportional to A/B_Π,

(v) a constant α_Π defined by

$$\alpha_\Pi = 4 \sum_{\Sigma\text{-states}} \frac{\langle \Pi | AL_x + 2B_x | \Sigma \rangle \langle \Sigma | BL_x | \Pi \rangle}{E_\Sigma - E_\Pi} \tag{IV.37}$$

(vi) another constant β_Π defined by

$$\beta_\Pi = 4 \sum_{\Sigma\text{-states}} \frac{|\langle \Pi | BL_x | \Sigma \rangle|^2}{E_\Sigma - E_\Pi}. \tag{IV.38}$$

Hence the analysis of the Λ-doubling frequencies yields values for the constants α_Π and β_Π, and also the rotational distortion parameter. Dousmanis, Sanders and Townes (1955) and Radford (1961, 1962) obtain the values

$$A/B_\Pi = -7{\cdot}504, \qquad \alpha_\Pi = -2361{\cdot}37 \text{ MHz}, \qquad \beta_\Pi = 576{\cdot}18 \text{ MHz}.$$

A good electronic wave function for OH should be capable of providing theoretical values of α_Π and β_Π which agree with the measured ones.

Next we consider the magnetic hyperfine interaction with the proton spin which results in a doublet splitting of each Λ-doublet component. The physics of the hyperfine coupling was described in section III.2.5 and in OH all the magnetic interactions discussed earlier are important. We recall that there are three different interactions, namely, the electron spin–nuclear spin dipolar interaction, the Fermi contact interaction, and the electron orbital–nuclear spin interaction. The first comprehensive discussion of these couplings was given by Frosch and Foley (1952) who showed that the complete magnetic hyperfine Hamiltonian can be expressed in terms of molecule-fixed components in the following form:

$$
\begin{aligned}
\mathcal{H}_{\text{hfs}} = {} & a I_z L_z && \text{(i)} \\
& + \tfrac{1}{2}a\{I_+ L_- + I_- L_+\} && \text{(ii)} \\
& + (b + c)I_z S_z && \text{(iii)} \\
& + \tfrac{1}{2}b\{I_+ S_- + I_- S_+\} && \text{(iv)} \\
& + \tfrac{1}{2}d\{e^{2i\phi}I_- S_- + e^{-2i\phi}I_+ S_+\} && \text{(v)} \\
& + e\{e^{i\phi}(S_- I_z + I_- S_z) + e^{-i\phi}(S_+ I_z + I_+ S_z)\} && \text{(vi)}
\end{aligned}
$$

$$\text{(IV.39)}$$

The operators $I_+, I_-, L_+, L_-, S_+, S_-$ are called shift operators, defined by

$$I_+ = I_x + i I_y, \qquad I_- = I_x - i I_y, \quad \text{etc.}$$

Terms (i) and (ii) come from the electron orbital–nuclear spin interaction, the constant b in terms (iii) and (iv) contains contributions from both the Fermi contact and spin-spin dipolar interactions and terms (v) and (vi) arise from the spin-spin dipolar coupling alone. Frosch and Foley give explicit expressions for the five constants a, b, c, d and e in terms of quantities describing the electronic wave function. Figure IV.22 shows the coordinate system used to describe the relative positions of the electron (e) and the proton (H), and the constants are expressed in terms of the coordinates $\mathbf{r}$, θ, ϕ as follows:

$$a = 2\mu_B\mu_N g_I \langle 1/r^3 \rangle$$

$$b = -\mu_B\mu_N g_I \langle (3\cos^2\theta - 1)/r^3 \rangle + (16\pi/3)\mu_B\mu_N g_I \psi^2(0)$$

$$c = 3\mu_B\mu_N g_I \langle (3\cos^2\theta - 1)/r^3 \rangle \qquad\qquad\text{(IV.40)}$$

$$d = 3\mu_B\mu_N g_I \langle \sin^2\theta/r^3 \rangle$$

$$e = 3\mu_B\mu_N g_I \sin\theta\cos\theta/r^3.$$

Of course the electron is not localized at one position in the molecule but has a probability distribution which can be calculated from the electronic wave function. Consequently the hyperfine constants depend upon averages over the electronic wave function, denoted by the symbol $\langle\ \rangle$ in equations (IV.40). (The term containing e mixes different electronic states.) In addition it is possible to determine $\psi^2(0)$, the density of electron spin at the proton nucleus.

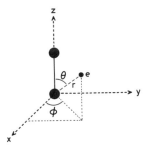

Fig. IV.22. Coordinate system used to describe the hyperfine interaction between an electron (e) and a nucleus in a diatomic molecule.

Clearly if it is possible to determine all five constants, a detailed description of the electronic wave function is provided. What, then, are the experimentally observable consequences of the six terms given in equation (IV.39)? They are as follows:

Terms (i) and (iii): These are diagonal in Λ and Σ and give identical splittings for both Λ-doublet components in any given J level.

Term (iv): This mixes the fine structure states $^2\Pi_{\frac{3}{2}}$ and $^2\Pi_{\frac{1}{2}}$.

Term (v): This term affects the two Λ-doublet components in a given J level, making equal but opposite contributions to the hyperfine splitting of each Λ-doublet component.

Terms (ii) and (vi): These mix different electronic states.

The hyperfine splittings are J dependent, and since terms (i), (iii) and (v) make important contributions to the splittings, the constants a, b, c and d can be determined provided sufficient measurements are available. By utilizing the pure microwave results of Dousmanis, Sanders and Townes and his own electron resonance results, Radford (1962) has given the following values of the proton constants in OH:

$$a = 86{\cdot}0 \text{ MHz}, \qquad b = -119{\cdot}0, \qquad c = 133{\cdot}2, \qquad d = 56{\cdot}5.$$

From these results we obtain values for the averaged quantities appearing in (IV.40), except for that involved in the constant e, which is not determined. Hence an accurate electronic wave function for OH must be able to account for these values, in addition to those of the Λ-doubling constants α_Π and β_Π presented earlier.

We must now consider the Zeeman effect which has been studied extensively by Radford. Application of an external magnetic field removes the $2J + 1$ degeneracy (neglecting the nuclear hyperfine splitting) and the rate at which the M_J levels diverge is determined by the appropriate g value for the J level in question. The Zeeman effect was discussed previously in section III.2.6 where we saw that the Zeeman Hamiltonian can be represented as the sum of four contributions,

$$\mathcal{H}_Z = \mu_B \mathbf{B} \cdot (g_L \mathbf{L} + g_e \mathbf{S} - g_I \mathbf{I} - g_r^n \mathbf{R}). \qquad \text{(IV.41)}$$

The first term is the orbital Zeeman interaction with $g_L = 1$, the second term is the electron spin Zeeman interaction with $g_e = 2{\cdot}00232$, the third term is the nuclear spin Zeeman interaction in which g_I for the proton has the value $3{\cdot}042 \times 10^{-3}$ (in units of the *electron* Bohr magneton μ_B) whilst the last term is the rotational Zeeman effect arising from the angular momentum $\mathbf{R}$ of the nuclei only. A classical calculation of g_r^n for OH gives a value $5{\cdot}42 \times 10^{-4}$.

Now as Radford points out, the Zeeman effect for OH is predominantly linear (first-order) and the Zeeman Hamiltonian (IV.41) can be rewritten in the form

$$\mathcal{H}_Z = \mu_B \mathbf{B} \cdot (g_J^{op} \mathbf{J} - g_I \mathbf{I}) \qquad \text{(IV.42)}$$

in which the effects of the terms containing $\mathbf{L}$, $\mathbf{S}$ and $\mathbf{R}$ are incorporated into the first term of (IV.42), the g_J^{op} being an operator which has the form

$$g_J^{op} = [J(J + 1)]^{-1}\{(g_L + g_r^n)(J_x L_x + J_y L_y + J_z L_z)$$
$$+ (g_e + g_r^n)(J_x S_x + J_y S_y + J_z S_z) + g_r^n \mathbf{J}^2\}. \qquad \text{(IV.43)}$$

If the matrix elements of this operator are calculated, one then obtains expressions for effective g factors of both Λ-doublet components of a given J level. These expressions are

$$g_J^+ = g_J^0 + (\delta g_J)_S + (\delta g_J)_R + (\delta g_J)_L^+,$$
$$g_J^- = g_J^0 + (\delta g_J)_S + (\delta g_J)_R + (\delta g_J)_L^-.$$

(IV.44)

where the $+$ and $-$ superscripts refer to the symmetric and anti-symmetric Λ-doublet components. These expressions for the g factors are useful because they permit a ready breakdown of the different contributions to the overall Zeeman effect. g_J^0 is the contribution arising from the operator $\mu_B(\mathbf{L} + 2\mathbf{S})\mathbf{B}$ acting on wave functions which include the rotational decoupling of $\mathbf{S}$ but not the effects of $\mathbf{L}$ uncoupling brought about by admixture of excited electronic states with the ground state. $(\delta g_J)_S$ represents the Zeeman correction due to the radiative correction (0·00232) to the spin g factor. $(\delta g_J)_R$ is the contribution due to the rotating nuclei, and the final contributions, $(\delta g_J)_L^+$ and $(\delta g_J)_L^-$ are corrections arising from $\mathbf{L}$ uncoupling. These final corrections are different for the $(+)$ and $(-)$ Λ-doublet levels and therefore lead to the overall g_J^+ and g_J^- factors being different. Radford obtained the following results for a number of different rotational levels of both the $^2\Pi_{\frac{3}{2}}$ and $^2\Pi_{\frac{1}{2}}$ states in OH:

$^2\Pi_{\frac{3}{2}}$	$J = \frac{3}{2}$	$g_J^+ = 0\cdot93493$	$g_J^- = 0\cdot93622$
	$J = \frac{5}{2}$	$0\cdot48435$	$0\cdot48623$
	$J = \frac{7}{2}$	$0\cdot32454$	$0\cdot32668$
$^2\Pi_{\frac{1}{2}}$	$J = \frac{3}{2}$	$-0\cdot13344$	$-0\cdot13443$
	$J = \frac{5}{2}$	$-0\cdot14092$	$-0\cdot14134$

Note that the g values for the $^2\Pi_{\frac{1}{2}}$ state are relatively small; we shall see later that a "pure" $^2\Pi_{\frac{1}{2}}$ state is essentially non-magnetic ($g_J = 0$) but in OH the rotational mixing of the fine structure states confers a substantial magnetic moment on the molecule in the $^2\Pi_{\frac{1}{2}}$ state. Needless to say, the above values offer a further challenge to the calculator of electronic wave functions.

There is a small second-order Zeeman effect in OH, similar in its effects to that described previously for $^1\Delta$ molecules. The magnetic field mixes different J levels with the result that the Zeeman splittings are not exactly linear and consequently the three $\Delta M_J = \pm 1$ transitions occur at slightly different field values. This is the reason for the triplet splitting in the $J = \frac{3}{2}$, $^2\Pi_{\frac{3}{2}}$ spectrum illustrated in Fig. IV.21(b). Analysis of the splitting yields a value for the rotational constant; however we shall see that this aspect of electron resonance spectra is much more important in heavier molecules where the rotational levels are closer together and therefore more strongly mixed by the applied magnetic field.

Other electron resonance investigations of the OH radical have been carried out. Carrington and Lucas (1970) have examined the spectrum of ^{17}OH artificially enriched in ^{17}O which has a spin of $\frac{5}{2}$. There is additional magnetic hyperfine splitting from ^{17}O and also an electric quadrupole contribution from which the ^{17}O quadrupole coupling constant is determined. The values obtained are encouragingly close to those calculated from *ab initio* wave functions. Interesting studies of vibrationally excited OH have also been described by Clough, Curran and Thrush (1971), Churg and Levy (1970), and Lee, Tam, Larouche and Woonton (1971). If the OH is produced by reaction of H atoms with O_3, the mixing occurring inside the electron resonance cavity, spectra arising from molecules in at least the $v = 4$ level are observed. Virtually all the molecular constants exhibit a marked dependence on the vibrational quantum number. A satisfactory interpretation has been provided recently by Hinkley, Walker and Richards (1973).

OH certainly qualifies as the free radical most studied in the laboratory and is a good example of the way in which a molecular wave function can be probed by high resolution spectroscopy. This radical also happens to be an important occupant of interstellar gas clouds, however, and is now being studied by many groups of astronomers in different parts of the world. In Chapter II we outlined very briefly the techniques used by astronomers to study radio and microwave emission from interstellar space, and OH is certainly keeping many telescopes and even more brains quite busy. The number of discrete sources detected now runs into hundreds, several rotational levels have been observed, and spectra arising from spontaneous emission, stimulated emission and absorption have been recorded. Of these, the stimulated emission lines are the most exciting and puzzling, since they indicate that population inversion of the Λ-doublet components occurs. Several explanations of this have been proposed, physical and chemical, and it may well be that different mechanisms operate in different sources. There is certainly no doubt that these spectra can tell us much about the physics and chemistry of interstellar gas clouds, once the code has been cracked. This must surely be one of the most fascinating areas of contemporary science.

A number of other diatomic hydrides with $^2\Pi_{\frac{3}{2}}$ ground states have been investigated, but less thoroughly than OH. Radford (1964a) showed that by mixing discharged water vapour with H_2S, H_2Se or H_2Te, he could detect the electron resonance spectra of SH, SeH or TeH respectively. Since the fine structure splitting increases and the rotational constant decreases from OH to TeH, case (a) coupling provides an increasingly accurate description of the ground states of these radicals and the Λ-doubling becomes progressively smaller. Indeed it is only just resolved in the $J = \frac{3}{2}$, $^2\Pi_{\frac{3}{2}}$ spectrum of TeH. Carrington, Levy and Miller (1967b) have studied the Stark effect in SH and the spectrum of the $J = \frac{3}{2}$ level shown in Fig. IV.23 illustrates an interesting feature. The top spectrum is recorded using magnetic field modulation with no Stark field. Ignoring the central line which is actually due to molecular oxygen, we see that the spectrum consists of two groups of six lines. The separation between the two groups is due to the Λ-doubling, whilst the six lines arise from the second-order Zeeman triplet splitting combined with the proton doublet splitting. The spectra shown in Fig. IV.23(b) and (c), however, are obtained using electric field modulation and a static electric field. The most striking feature is the appearance of new lines approximately midway between the main groups; the transitions giving rise to these lines would, in the absence of an applied electric field, have magnetic dipole intensity only because they occur between states of the same parity. However we have already seen that an electric field mixes states of opposite parity, so that in the presence of an applied electric field, the states no longer have definite parity. Consequently the transitions now acquire some electric dipole intensity and are therefore readily observable. The details of the observed line shapes are discussed in the original paper. The changes in the spectrum also allow the electric dipole moment to be determined. Subsequently Miller (1971b) has detected the satellite lines arising from naturally-abundant ^{33}S and determined some of the hyperfine constants. The spectra of SeH and SeD have also been investigated in detail by Carrington, Currie and Lucas (1970); they were able to determine the electric dipole moment through studying the Stark effect, and its value makes an interesting comparison with those of OH and SH:

$$\mu(\text{OH}) = 1\cdot66 \text{ D}, \qquad \mu(\text{SH}) = 0\cdot62 \text{ D}, \qquad \mu(\text{SeH}) = 0\cdot46 \text{ D}.$$

There are two final points worth noting about SH, SeH and TeH. Firstly, attempts to detect the pure microwave spectra of these radicals have so far failed; consequently the effective Hamiltonian parameters are rather inadequately determined at present, even for SH. Secondly, because rotational distortion of the ground states is so small compared with that in OH, the $^2\Pi_{\frac{1}{2}}$ states are essentially non-magnetic and therefore not amenable to electron resonance investigations. It is not difficult to show that for a good case (a) molecule the g value for the rotational level J is given by

$$g_J = \frac{(\Lambda + \Sigma)(\Lambda + g_e\Sigma)}{J(J + 1)}. \tag{IV.45}$$

For the $J = \frac{3}{2}$ rotational level of a $^2\Pi_{\frac{3}{2}}$ state and putting $g_e = 2$, (IV.45) gives the result $g_J = \frac{4}{5}$. For any rotation level of the $^2\Pi_{\frac{1}{2}}$ state, however, (IV.45)

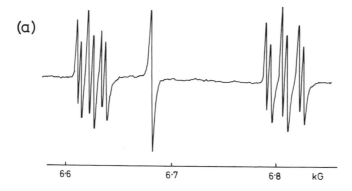

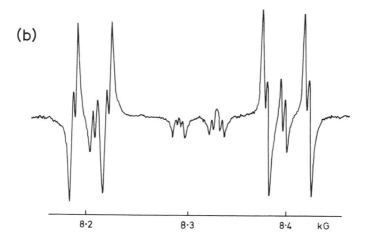

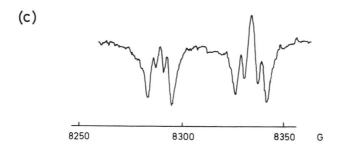

Fig. IV.23. *X*-band electron resonance spectra of SH($J = \frac{3}{2}$, $^2\Pi_{\frac{3}{2}}$). (a) Magnetic field modulation. (b) Electric field modulation. (c) Electric-field induced transitions.

gives $g_J = 0$. Physically the magnetic moments due to **L** and **S** which, in a good case (a) molecule, lie along the molecular axis add together in the $^2\Pi_{\frac{3}{2}}$ state but are opposed and effectively cancel each other in the $^2\Pi_{\frac{1}{2}}$ state. For a molecule like OH where the fine structure states are mixed by the nuclear rotation, the magnetic moment due to **S** no longer lies exactly along the molecular axis. Consequently both fine structure states possess substantial magnetic moments, their relative sizes depending on the strength of the rotational distortion.

In conclusion we should say something about the CH radical, which has been sought by microwave spectroscopists of all sorts without success, but which has recently succumbed to the laser electron resonance method of Evenson, Radford and Moran (1971). Their experimental technique was described in detail in section II.2.6. The CH radicals are generated inside the sample cell by means of a low pressure oxyacetylene flame and the radiation source is a water vapour laser operating at a frequency of 2527·95 GHz. The energy levels are tuned into resonance with an applied magnetic field and the transitions detected are Zeeman components of the $J = \frac{5}{2} \to \frac{7}{2}$ rotational

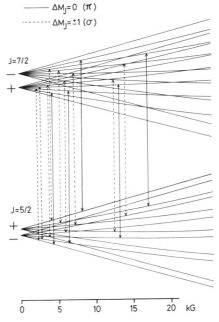

Fig. IV.24. Zeeman and Λ-doublet components of the $J = \frac{5}{2}$ to $J = \frac{7}{2}$ rotational transitions in CH, studied by laser electron resonance (Evenson, Radford and Moran, 1971).

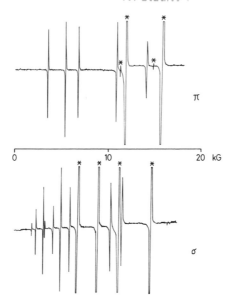

Fig. IV.25. Laser electron resonance spectrum of CH. The lines marked with an asterisk are due to OH (Evenson, Radford and Moran, 1971).

transition in the $^2\Pi_{\frac{1}{2}}$ ground state. The appropriate energy level diagram is shown in Fig. IV.24 and the electron resonance spectrum in Fig. IV.25. The $\Delta M_J = \pm 1$ transitions (σ-type) are induced when the electric vector of the laser radiation is perpendicular to the applied magnetic field, whilst the $\Delta M_J = 0$ (π-type) transitions correspond to parallel polarization. From a preliminary analysis of the data, values for the rotational constant (14·162 cm^{-1}), rotational distortion constant ($A/B = 1\cdot99$) and difference in Λ-doubling separations for the two levels ($v_\Lambda(\frac{7}{2}) - v_\Lambda(\frac{5}{2}) = 0\cdot213$ cm^{-1}) are obtained. The line width observed is (in frequency units) 21·5 MHz which is, of course, one or two orders-of-magnitude larger than would be expected with a microwave spectrum. It will be interesting to see if measurements of the Λ-doubling in the lowest ($J = \frac{1}{2}$) rotational level can be made, since astronomers have a particular interest in the detection of interstellar CH, but lack precise knowledge of the relevant transition frequencies at present.

IV.6.3. Molecules with small Λ-doubling

A considerable number of diatomic radicals in $^2\Pi_{\frac{3}{2}}$ states have now been detected and studied, particularly through electron resonance methods.

Apart from the hydrides discussed in the previous section, these at present include

NO, ClO, BrO, IO, CF, SF, SeF, NS, LiO

The main difference between these species and the hydrides is that, apart from NO and LiO, the Λ-doubling in the rotational levels studied (usually only the first one or two) is unresolved. Moreover the fine structure splittings are larger and the rotational constants are smaller, so that most of the radicals conform fairly closely to Hund's case (a) coupling. Indeed this is one of the main reasons why these species have been successfully detected, since the predicted g_J value of $\frac{4}{5}$ for the lowest rotational level ($J = \frac{3}{2}$) gives an excellent indication of the magnetic field range to be examined in an electron resonance experiment.

Since the general principles governing the energy levels of these species are similar in all cases, we shall confine detailed attention to two or three molecules only. Apart from NO and LiO, the ClO radical has been studied more thoroughly than any of the others listed above, since it has been detected by pure microwave spectroscopy as well as by electron resonance. The first investigations were by Carrington, Dyer and Levy (1967), using electron resonance methods, who found that the $J = \frac{3}{2}$ ClO spectrum could be detected by passing the products of a microwave discharge in chlorine/oxygen mixtures through the electron resonance cavity. The spectrum they obtained is shown in Fig. IV.26(a) and it may be readily understood in terms of the energy level diagram shown in Fig. IV.27. The ultraviolet spectrum had already established the ground state to be $^2\Pi_{\frac{3}{2}}$, the fine structure splitting being about 280 cm^{-1}. In the lowest rotational level ($J = \frac{3}{2}$) the applied magnetic field removes the $2J + 1$ degeneracy and three $\Delta M_J = \pm 1$ lines arise because of transitions induced by the electric vector of the microwave field between the components of opposite parity. Each level is two-fold degenerate because of the effective Λ-degeneracy, any Λ-doublet splitting being smaller than the line widths in the microwave spectra. The three $\Delta M_J = \pm 1$ lines occur at different field strengths because of the second-order Zeeman effect which mixes the $J = \frac{3}{2}$ level with $J = \frac{5}{2}$. From an analysis of this splitting, which results in the three separated groups in Fig. IV.26(a), the rotational constant B_0 was determined. There are two isotopes of chlorine with nuclear spin, ^{35}Cl and ^{37}Cl, both with $I = \frac{3}{2}$ and in a natural abundance ratio of approximately 3:1, so that the strongest hyperfine components arise from ^{35}Cl. The chlorine quartet splitting is clearly seen in the spectrum, and arises from the magnetic hyperfine interactions discussed in the previous section. We note, however, that the lines are evenly spaced in the centre group, but have unequal spacings in the outer groups. This is due to the

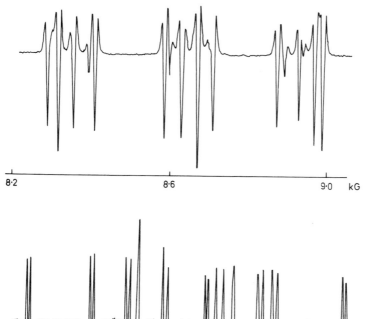

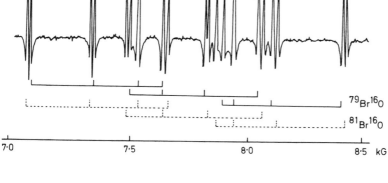

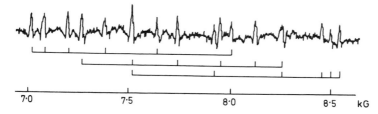

Fig. IV.26. X-band electron resonance spectra of (a) ClO ($J = \frac{3}{2}$, $^2\Pi_{\frac{3}{2}}$), (b) BrO ($J = \frac{3}{2}$, $^2\Pi_{\frac{3}{2}}$), (c) IO ($J = \frac{3}{2}$, $^2\Pi_{\frac{3}{2}}$).

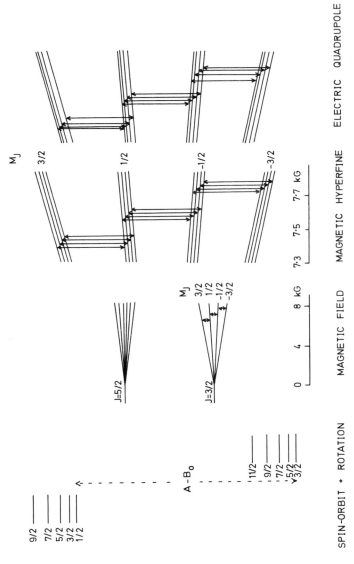

Fig. IV.27. Energy level diagram showing the interpretation of the electron resonance spectrum of ClO.

nuclear electric quadrupole interaction and a straightforward analysis yields both the magnitude and sign of the chlorine quadrupole coupling constant.

Subsequent to the electron resonance investigations, the pure microwave spectrum was studied by Amano, Saito, Hirota, Morino, Johnson and Powell (1969). They measured the $J = \frac{3}{2} \to \frac{5}{2}$ rotational transition in the $^2\Pi_{\frac{3}{2}}$ ground state (the hyperfine components of which span the frequency range 91–93 GHz) and the $J = \frac{1}{2} \to \frac{3}{2}$ rotational transitions in the $^2\Pi_{\frac{1}{2}}$ state (54·5–56·5 GHz). The appropriate energy level diagram and transitions for the $^2\Pi_{\frac{3}{2}}$ state are shown in Fig. IV.28. By combining the electron resonance and pure microwave results, it has proved possible to determine a number of the molecular constants. The rotational constants for both isotopic species are determined most accurately from the pure microwave measurements, and the centrifugal distortion parameters have also been measured. The magnetic hyperfine constants are more difficult to unravel in a good case (a) molecule because the magnetic moments due to $\mathbf{L}$ and $\mathbf{S}$ lie rather accurately along the molecular axis. Measurements in the $^2\Pi_{\frac{3}{2}}$ state yield the value of $a + \frac{1}{2}(b + c)$, whilst in the $^2\Pi_{\frac{1}{2}}$ state, where the moments due to $\mathbf{L}$ and $\mathbf{S}$ are opposed, the axial hyperfine component is $a - \frac{1}{2}(b + c)$. Consequently measurements in both fine structure states yield the value of the a hyperfine constant directly. The b hyperfine constant is more difficult to determine because it is related to the rotational distortion of the $^2\Pi$ state, which in turn depends on the fine structure splitting (A) which is, unfortunately, not known accurately from the electronic spectrum. The problem is that variations in the values of A and b affect the predicted microwave and electron resonance spectra in similar

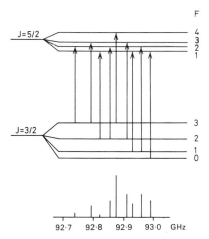

Fig. IV.28. Hyperfine components of the $J = \frac{3}{2}$ to $J = \frac{5}{2}$ rotational transition in ClO.

ways, although the hyperfine mixing of the fine structure states is usually small compared with the rotational mixing. Amano *et al.* conclude that the experimental measurements are best fitted with A equal to $-282 \, \text{cm}^{-1}$, and b equal to 63 MHz. The quadrupole coupling constant and the electric dipole moment (1·23 D) have been determined from both the electron resonance and microwave experiments with good agreement.

Figure IV.26(b) and (c) shows the electron resonance $(J = \frac{3}{2})$ spectra of BrO and IO, detected by Carrington, Dyer and Levy (1970) who made use of the reactions of oxygen atoms with bromine and trifluoroiodomethane respectively. The structure of these spectra is essentially the same as that of ClO, even though they may at first sight appear different. The spectrum of BrO is complicated by the fact that the isotopic species ^{79}BrO and ^{81}BrO are present in virtually equal abundance, so that it is quite difficult to decide initially to which species a given line should be attributed. Furthermore the second-order Zeeman splitting is not immediately recognizable because the bromine hyperfine and quadrupole couplings are so large. Nevertheless the analysis given below the BrO spectrum shows that the basic structure is very much the same as that of the ClO spectrum, the nuclear spin of both bromine isotopes again being $\frac{3}{2}$. The spectrum of IO shown in Fig. IV.27(c) is again of the same form, the main difference being that ^{127}I is the only important isotope and it has a spin of $\frac{5}{2}$, with a consequent sextet hyperfine splitting.

The analysis of the BrO and IO spectra follows along much the same lines as that of ClO, except for one important new feature. The experimental g_J value depends, among other things, on the extent of the rotational mixing of the $^{2}\Pi_{\frac{3}{2}}$ and $^{2}\Pi_{\frac{1}{2}}$ states. If the rotational constant is known, the observed g_J value can be used to provide an estimate of A, but this leads to the rather absurd result that the fine structure splitting in IO is smaller than in BrO. Clearly there is another contribution to the g_J value which is important in IO (and other heavy molecules) and this is, in fact, the spin-orbit mixing of the $^{2}\Pi$ ground state with excited electronic states (term (iv) in equation (IV.30)); Brown, Byfleet, Howard and Russell (1972) have measured the second rotational levels of BrO, IO and SeF and shown that satisfactory interpretations of the g_J values are obtained if this effect, which actually represents a transition from Hund's case (a) to case (c) coupling, is taken into account. They are also able to provide an estimate of the overall rotational g factor.

The other $^{2}\Pi$ radical whose electron resonance spectrum we illustrate is that of CF, shown in Fig. IV.29 which is taken from the paper by Carrington and Howard (1970). The $J = \frac{3}{2}$ spectrum, which was obtained (accidentally!) by mixing fluorine atoms with acetonitrile, is readily analysed in terms of the usual triplet Zeeman splitting and a doublet hyperfine splitting from ^{19}F. The $J = \frac{5}{2}$ spectrum consists of five doublets, since five $\Delta M_J = \pm 1$ transitions are

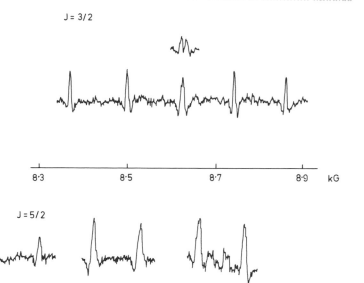

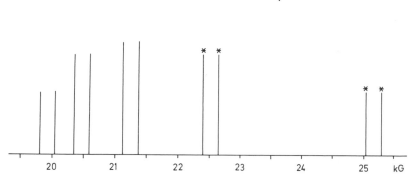

Fig. IV.29. Electron resonance spectra of CF ($^2\Pi_{\frac{3}{2}}$) in its $J = \frac{3}{2}$ and $J = \frac{5}{2}$ rotational
levels.

possible, although the two highest field doublets were beyond the magnetic
fields available to the authors. Apart from confirming values of B_0 and A
determined previously from the electronic spectrum, the electron resonance
spectrum provides a value for the axial hyperfine component $a + \frac{1}{2}(b + c)$,
and it is also possible to determine separately the value of b from the measurements of both rotational levels. Because B_0/A is relatively large the rotational
distortion is considerable and this, together with the fact that the fluorine
hyperfine interaction is large, means that the constant b can be estimated

fairly accurately. It should be noted that the $^2\Pi_{\frac{3}{2}}$ state is not, in fact, the ground state since A is positive (as in CH) and the ground $^2\Pi_{\frac{1}{2}}$ state is 77 cm^{-1} lower in energy.

Carrington and Howard also studied the Stark effect, which is essentially first-order because the Λ-doublets are effectively degenerate (as they are in most of the other species considered in this section). They determined the electric dipole moment to be 0·65 D, and it is interesting to note that calculations of the molecular wave function by O'Hare and Wahl (1971) lead to a dipole moment in good agreement with that measured, but indicate the polarity to be C$^-$F$^+$, which might well surprise some readers.

Finally we must discuss the recent work of Freund, Herbst, Mariella and Klemperer (1972) who have succeeded in carrying out the first molecular beam electric resonance study of a ground state free radical, namely LiO. The details of their experimental method were described in Chapter II and although these are quite different from the pure microwave and electron resonance experiments, the spectroscopy is of exactly the same kind, the transitions detected again being between Λ-doublet components of a given J level. Although the Λ-doubling in the lowest rotational level of LiO is quite small (~ 11 MHz), Freund *et al.* were actually able to study a considerable number of higher rotational levels, the Λ-doubling increasing up to ~ 12 GHz. Consequently the analysis of the spectra is more closely related to that of OH and OD than to the other molecules described in this section.

Freund, Herbst, Mariella and Klemperer concentrated on the spectrum of the isotopic species ^{7}Li^{16}O in which ^{7}Li has a nuclear spin of $\frac{3}{2}$. The fine

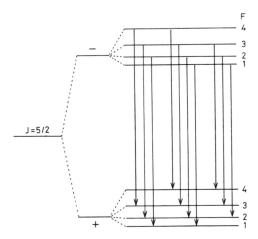

Fig. IV.30. Energy levels and electric resonance transitions in the $J = \frac{5}{2}$ ($^2\Pi_{\frac{3}{2}}$) rotational level of LiO (Freund, Herbst, Mariella and Klemperer, 1972).

structure separation is about 112 cm^{-1}, the ground state being the $^2\Pi_{\frac{3}{2}}$ component. The transitions studied were between Λ-doublet components, split by hyperfine interaction and obeying the selection rules $\Delta J = 0$, $\Delta F = 0, \pm 1$. Figure IV.30 shows, for example, the energy levels and transitions in the $J = \frac{5}{2}$ level of the $^2\Pi_{\frac{3}{2}}$ state. Lines from $J = \frac{3}{2}$ (~ 11 MHz frequency) to $J = \frac{25}{2}$ (~ 3873 MHz) in the $^2\Pi_{\frac{3}{2}}$ component, and from the $J = \frac{3}{2}$ level of the $^2\Pi_{\frac{1}{2}}$ component ($\sim 12{,}320$ MHz) were measured. Some additional lines from the $v = 1$ and 2 excited vibrational levels were also observed. Compared with the electron resonance and pure microwave investigations of other radicals, this represents a very complete study, as a result of which it was possible to determine the values of the hyperfine constants a, b, c and d (see equation (IV.40)), the lithium quadrupole coupling constant, and the Λ-doubling parameter α_Π. Furthermore the observed Stark effect yielded a value for the electric dipole moment (6·84 D), which agrees very well with a theoretical value calculated by Wahl (1972). The results are consistent with the simplest view of LiO as an essentially ionic molecule, Li$^+$O$^-$.

IV.7. $^3\Sigma$ States

A number of $^3\Sigma$ states have now been studied, including O_2, SO and SeO in their ground states and, more recently, N_2 in its $A\ ^3\Sigma_u^+$ state which lies approximately 50,000 cm^{-1} above the $^1\Sigma_g^+$ ground state. In the absence of hyperfine and external field interactions the effective Hamiltonian contains terms describing the nuclear rotation, the electron spin-spin interaction, and the electron spin–nuclear rotation interaction, i.e. for the $v = 0$ level,

$$\mathscr{H} = B_0 \mathbf{N}^2 + \tfrac{2}{3}\lambda_0(3S_z^2 - \mathbf{S}^2) + \gamma_0 \mathbf{N}.\mathbf{S}. \qquad (\text{IV.46})$$

The first and third terms are familiar from our previous discussions of other states, but the second term requires more discussion. We need not dwell on the precise form of this term except to note that it is quadratic in the electron spin operator $\mathbf{S}$. This is not unexpected if the term arises from the dipole-dipole interaction of the two unpaired electron spins. However we recall from section III.2.7 that spin-orbit coupling can also give rise to a term of the same form. Although a "pure" $^3\Sigma$ state has no first-order spin-orbit coupling ($\Lambda = 0$), we noted that particularly in molecules containing heavy atoms, the spin-orbit interaction can mix the $^3\Sigma$ state with other electronic states of different spin multiplicity or different Λ value. In other words, we are not dealing with a "pure" $^3\Sigma$ state, and it turns out that the observable effects of this spin-orbit coupling are the same as those of the direct dipolar interaction between the unpaired spins in the ground state. Hence the so-called spin-spin constant λ_0 in (IV.46) consists of a first and second-order contribution,

$$\lambda_0 = \lambda^{(1)} + \lambda^{(2)} \qquad\qquad\text{(IV.47)}$$

where the first-order part $\lambda^{(1)}$ does indeed represent the direct spin-spin interaction, but the second-order part $\lambda^{(2)}$ arises from the indirect spin-orbit coupling. Even in a relatively light molecule like O_2 in its $^3\Sigma_g^-$ ground state ($\lambda_0 = 1\cdot985\,\text{cm}^{-1}$), $\lambda^{(2)}$ is comparable with $\lambda^{(1)}$ and in a fairly heavy molecule like SeO ($\lambda_0 = 86\cdot4\,\text{cm}^{-1}$), $\lambda^{(2)}$ completely dominates $\lambda^{(1)}$. So although we shall continue to refer to λ_0 as the spin-spin constant, it is as well to remember that this description is consistent with the form of the term in the effective Hamiltonian, rather than with the molecular physics which are responsible for that term.

The form of the rotational levels of a $^3\Sigma$ state very much depends upon the relative magnitudes of B_0 and λ_0. If λ_0 is small, the coupling approximates closely to Hund's case (b) as shown on the left-hand side of Fig. IV.31. Each rotational level $N = 0, 1, 2, 3$, etc. is split into a triplet by the second and third terms in (IV.46), the final levels being characterized by the values of J,

$$J = N + 1, N, N - 1,$$

except for the $N = 0$ level where only the $J = 1$ component exists. Now as λ_0

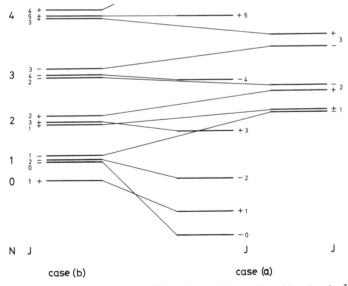

Fig. IV.31. Correlation of the case (b) and case (a) rotational levels of a $^3\Sigma$ state.

becomes larger and B_0 smaller, the coupling approaches case (a), as shown on the right-hand side of Fig. IV.31. It is convenient to arrange the case (a) levels in two different stacks as shown, one consisting of widely spaced single levels, and the other consisting of pairs of levels which are nearly degenerate for low J values. Of the molecules which have been studied by high-resolution techniques, O_2 and N_2 ($A\ ^3\Sigma_u^+$) are intermediate between case (b) and case (a), but SO and especially SeO are much closer to case (a) coupling.

Molecular oxygen has been studied extensively by microwave techniques, particularly by Tinkham and Strandberg (1955), but it differs from the other ground state $^3\Sigma$ species we shall discuss in that, since it does not possess an electric dipole moment, the transitions are necessarily of magnetic dipole intensity only. Hundreds of microwave and electron resonance lines are observable, however, because (since O_2 is a stable gas) the concentration in the sample cell is sufficient for the observation of the magnetic dipole spectrum. The pure microwave spectrum arises predominantly from $\Delta J = \pm 1$, $\Delta N = 0$ transitions which mainly occur in the frequency range 53 to 65 GHz. The $J = N \pm 1$ levels occur as closely spaced pairs, well separated from the $J = N$ component. (Note also that because of symmetry restrictions only odd N values are allowed.) Consequently the $\Delta J = \pm 1, \Delta N = 0$ transitions occur at relatively high frequencies. In a magnetic field, however, the $J = N \pm 1$ levels are mixed together with the result that the previously forbidden $\Delta J = \pm 2$ transitions can now be studied in a resonance experiment and occur at relatively low microwave frequencies because of the proximity of the $J = N \pm 1$ levels. These magnetic resonance transitions satisfy the additional selection rule $\Delta M_J = \pm 1$. In more recent work by Evenson $et\,al.$ (1968) using the laser magnetic resonance method at a frequency of 891 GHz, a $\Delta N = \pm 1$, $\Delta J = +1, \Delta M_J = 0$ transition has also been studied.

Many papers have now appeared discussing the theory and interpretation of the microwave measurements on O_2 in its various isotopic and hyperfine modifications. The discussion of Tinkham and Strandberg (1955) of the pure microwave spectrum is concerned mainly with corrections to the Born-Oppenheimer approximation and the related vibrational dependence of the spin-spin constant λ. Their discussion of the magnetic resonance spectrum deals mainly with the determination of the rotational magnetic moment (or rotational g factor g_r) and the g_i^e factor arising from the spin-orbit mixing of excited electronic states with the ground state. The theoretical analysis of the O_2 results represents a prototype for the analysis of other $^3\Sigma$ spectra; it is, however, atypical because it is a magnetic dipole spectrum whereas the related radicals SO and SeO have been studied through their electric dipole spectra.

SO has been studied by pure microwave spectroscopy and by electron resonance techniques; it is a relatively long-lived free radical which can be

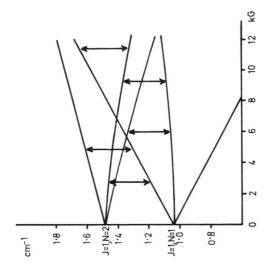

(b)

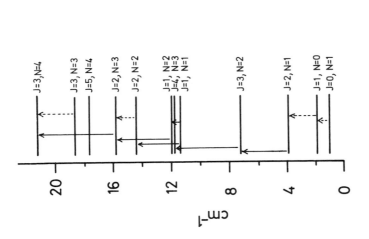

(a)

Fig. IV.32. (a) Lower rotational levels of SO and microwave transitions studied. (b) Electron resonance transitions in SO.

formed through a variety of chemical reactions. The lower rotational levels are illustrated in Fig. IV.32(a) and the microwave transitions detected are also shown. The solid lines indicate transitions studied by Winnewisser, Sastry, Cook and Gordy (1964) using the millimetre wave spectrometer described in Chapter II; the transition frequencies range from 86 GHz to 172 GHz. The dotted lines denote transitions studied by Powell and Lide (1964) using a Stark waveguide cell at frequencies ranging from 13 GHz to 66 GHz. Both groups of workers carried out essentially the same analysis with good agreement as to the values of the constant B_0, λ and γ. In addition Winnewisser et al. determined the centrifugal distortion constant D_0 to be 0·0334 MHz.

The electron resonance spectrum was studied by Daniels and Dorain (1966) and by Carrington, Levy and Miller (1967c). Both groups used a spectrometer operating at X-band frequencies and Daniels and Dorain also made additional measurements using an S-band instrument ($\sim 3\cdot2$ GHz). At these relatively low microwave frequencies it is necessary to find a pair of levels which are close in energy and Fig. IV.32(a) indicates that the most likely candidates are $J = 1$, $N = 1$ and $J = 1$, $N = 2$. An applied magnetic field removes the $2J + 1$ degeneracy of each level and Fig. IV.32(b) shows the splitting as a function of the field strength, together with the resonance transitions detected. Notice that all the transitions obey the selection rule $\Delta N = \pm 1$. If the pure microwave values of the constants B_0, λ and γ are used, it is possible to determine the Zeeman effect parameters from the electron resonance results. In particular the g_l^e and g_r factors were determined independently by both groups with good agreement. Lines from the isotopic species $^{34}S^{16}O$ and $^{33}S^{16}O$ were also detected, the latter arising from hyperfine interaction with ^{33}S. Furthermore lines from $^{32}S^{16}O$ in its first excited vibrational level were observed and from these Carrington, Levy and Miller were able to say something about the vibrational dependence of the spin-spin constant λ.

The SeO radical is similar in its electronic structure to SO, and it has been detected and studied by Carrington, Currie, Levy and Miller (1969). The main difference between SO and SeO, however, is that spin-orbit coupling is much more important in the latter, so that λ is much larger and the coupling tends strongly towards case (a). The transitions studied were again (in case (b) nomenclature) predominantly $J = 1$, $N = 1$ to $J = 1$, $N = 2$, the $2J + 1$ degeneracy being removed by the applied magnetic field as shown in Fig. IV.33. The four main transitions $\Delta M_J = \pm 1$ between states of opposite parity give rise to four lines but since the centre two lines occur at almost the same field value, the observed spectrum consists of three lines as shown in Fig. IV.33. However, in the presence of the modulating and bias electric fields the parity of the levels is not conserved, with the result that the "forbidden" transitions shown acquire electric dipole intensity and give rise to the weak lines shown in Fig. IV.33. The spectrum was interpreted using values of B_0 and λ determined from the ultraviolet spectrum and treating γ as a parameter to be determined by the best fit to the observed line positions. Hyperfine structure from ^{77}Se which has a spin of $\frac{1}{2}$ and a natural abundance of 7·5% was also observed and the axial component of the total (contact + dipolar) hyperfine interaction determined.

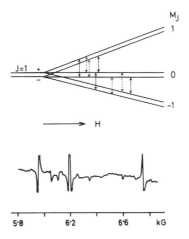

Fig. IV.33. Electron resonance transitions and spectrum for SeO in its $J = 1$ rotational level.

We conclude this discussion of $^3\Sigma$ states by describing recent work of Freund, Miller, De Santis and Lurio (1970) on the lowest excited electronic state ($A\ ^3\Sigma_u^+$) of N_2. The details of their molecular beam magnetic resonance experiments were described in Chapter II and we recall that a beam of N_2 is excited by electron impact into the $A\ ^3\Sigma_u^+$ state, which is metastable. The electron configurations of the ground ($X\ ^1\Sigma_g^+$) and lowest excited ($A\ ^3\Sigma_u^+$) states may be written in terms of molecular orbital theory as

$$X\ ^1\Sigma_g^+ : (1s\sigma_g)^2(1s\sigma_u^*)^2(2s\sigma_g)^2(2s\sigma_u^*)^2(2p\pi_u)^4(2p\sigma_g)^2$$

$$A\ ^3\Sigma_u^+ : \quad . \quad . \quad . \quad . \quad . \quad . \quad (2p\pi_u)^3(2p\sigma_g)^2(2p\pi_g^*)^1$$

The most remarkable feature of the work by Freund *et al.* is that they have detected spectra arising from excited molecules in the first thirteen vibrational levels ($v = 0$ to 12). For each vibrational level the zero-field states may be classified by their values of N, J and F. We discussed the nuclear spin functions of N_2 in section III.5.2 and noted that vectorial addition of the $I = 1$ spins of both ^{14}N nuclei resulted in a total nuclear spin I_T of 2, 1 or 0. We also showed that rotational levels of *even N* could only be associated with $I_T = 1$ and *odd N* with $I_T = 2$ and 0. Figure IV.34 shows (not to scale) the first few rotational levels and the vertical lines denote transitions studied in each of the first thirteen vibrational levels. In fact the magnetic resonance spectrum is obtained by using a magnetic field to remove the spatial degeneracy so that a simple zero-field $\Delta F = \pm 1$ transition is actually split into Zeeman subcomponents. Figure IV.35 shows, for example, the splitting in the $N = 2$, $J = 1$, $F = 2, 1, 0$ levels and the transitions detected by Freund *et al.* The $\Delta F = 1$ transitions are resolved for each vibrational level.

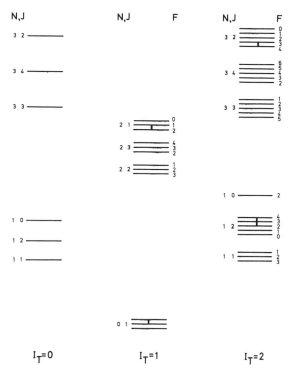

Fig. IV.34. Lower rotational levels of N_2 in its $A\ ^3\Sigma_u^+$ state, and the magnetic resonance transitions studied by Freund, Miller, De Santis and Lurio (1970).

The measured transition frequencies depend on the values of B, λ and γ, magnetic hyperfine constants describing the Fermi contact and dipolar interactions, the electric quadrupole coupling constant for ^{14}N and the magnetic coupling of the rotational and nuclear spin magnetic moments. Each of these constants is found to be vibrationally dependent, the variation being well represented by the usual power series expansion in $(v + \frac{1}{2})$.

Freund *et al.* reduce their data to values of averages over the electronic wave function, and consequently show how the wavefunction changes as, with increasing quanta of vibrational energy, the molecule approaches its dissociation limit. The spectra thus provide a detailed map of the electronic wavefunction covering more than half of the potential energy diagram for the $A\ ^3\Sigma_u^+$ state; this is a remarkable development in experimental spectroscopy, which probes more deeply than before some of the foundations on which our discussion of molecular structure rests at present.

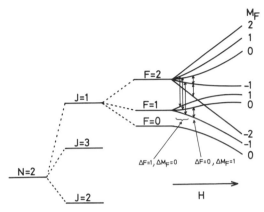

Fig. IV.35. Fine and hyperfine structure of the $N = 2$ rotational level of N_2 $(A\ ^3\Sigma_u^+)$ and the magnetic resonance transitions studied.

IV.8. $^3\Pi$ States

$^3\Pi$ states are of considerable interest and complexity, exhibiting many of the effects described previously for $^2\Pi$ and $^3\Sigma$ molecules. At the present time a ground $^3\Pi$ molecule has not been studied by high resolution techniques, but three studies of excited $^3\Pi$ states have been described. Two of these are concerned with two different excited $^3\Pi$ states of H_2, and the third refers to an excited $^3\Pi$ state of CO. Because of the very small spin-orbit coupling in H_2, the rotational levels and spectral analysis are most suitably discussed in terms of case (b) coupling, whereas in $^3\Pi$ CO case (a) coupling is more appropriate. We shall deal with these two cases separately.

The excited electronic states of H_2 were depicted earlier in Fig. IV.4. The lowest $^3\Pi_u$ state arises from excitation of one $1s\sigma$ electron into the $2p\pi$ level; it is named the $c\ ^3\Pi_u$ state, lies approximately 96,000 cm^{-1} above the ground state, and has been investigated by Lichten (1960) using the molecular beam magnetic resonance method discussed in Chapter II. The next $^3\Pi_u$ state is the

$d\ {}^3\Pi_u$ state lying about 113,000 cm^{-1} above the ground state, which has been studied recently by two groups working independently, Jost, Marechal and Lombardi (1972) and Miller and Freund (1972). Both groups used microwave or radiofrequency/optical double resonance techniques.

We describe first the work of Lichten who used electron bombardment to excite a beam of H_2 molecules into the $c\ {}^3\Pi_u$ state. The rotational levels of this state have been discussed by Fontana (1962) and Chiu (1966) using case (b) notation. Coupling of the component of orbital angular momentum along the internuclear axis (Λ) with the rotational angular momentum leads to **N**, the total angular momentum exclusive of electron spin. The rotational levels are characterized by values of $N = 1, 2, 3$, etc. and interaction of the electron spin **S** with **N** yields the total angular momentum **J**; each N level is therefore split into three spin components which are characterized by J values of $N + 1$, N and $N - 1$. Fontana investigated theoretically the energies of these levels, taking into account the nuclear rotation, the spin-spin interaction between the two unpaired electrons, and the spin-orbit coupling. The results of his theory are illustrated in Fig. IV.36, which shows the lower rotational levels in increasing order of energy. The most interesting feature is the variation of the fine structure splitting for each N level from para to ortho H_2. This variation has nothing to do with the proton nuclear spins, but arises from the fact that certain matrix elements of the electron spin-spin interaction make equal but opposite contributions to the energies of odd and even N levels. The levels shown in Fig. IV.36 retain the two-fold Λ-degeneracy, there being no direct evidence from the spectra of substantial Λ-doubling. The para H_2 levels, which have total nuclear spin I_T equal to zero, do not, of course, exhibit nuclear hyperfine splitting; the ortho H_2 levels which correspond to $I_T = 1$ do exhibit hyperfine splitting, however, which will be discussed later.

The first level studied in detail by Lichten was the $N = 2$ level of para H_2; Fig. IV.37 shows the three J levels involved and the splittings produced by an applied magnetic field. We recall from Chapter II that in a molecular beam magnetic resonance experiment the transitions are produced by an oscillating magnetic field in the presence of a static magnetic field (the C field). Lichten measured $\Delta M_J = \pm 1$, $\Delta J = \pm 1$ (π-type) and $\Delta M_J = 0$, $\Delta J = \pm 1$ (σ-type) transitions, part of his spectrum being illustrated in Fig. IV.37, and the results were then extrapolated by various means to zero magnetic field to yield the following values for the zero-field transition frequencies:

$$J = 2 \leftrightarrow J = 3, \qquad \nu = 5898{\cdot}105 \text{ MHz},$$

$$J = 1 \leftrightarrow J = 2, \qquad \nu = 4928{\cdot}003 \text{ MHz}.$$

Lichten was also able to measure the g factors for the levels involved.

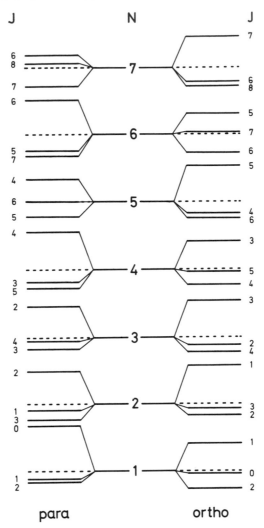

Fig. IV.36. Lower rotational levels of H_2 in its $c\ ^3\Pi_u$ state (Fontana, 1962).

In more recent work Brooks, Lichten and Reno (1971) have carried out similar measurements on the $N = 1$ level of ortho H_2, the fine and hyperfine structure of which is illustrated in Fig. IV.38. Note that each J component is split into a hyperfine triplet because ortho H_2 has $I_T = 1$. Brooks, Lichten and Reno measured a large number of transitions in the presence of a static magnetic field and although they were not able to identify positively all of

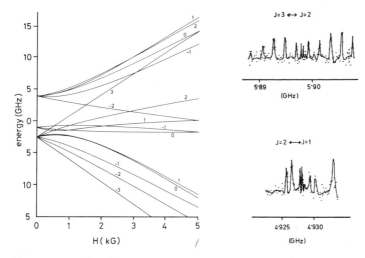

Fig. IV.37. Zeeman effect in the $N = 2$ rotational level of $c\ ^3\Pi_u$ para-H_2, and the magnetic resonance spectra detected by Lichten (1960).

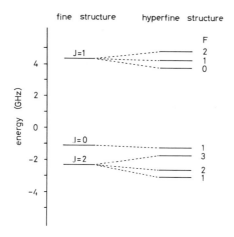

Fig. IV.38. Fine and hyperfine structure of the $N = 1$ rotational level in ortho-H_2 $(c\ ^3\Pi_u)$.

them, they were nevertheless able to establish the relative zero-field energies of all seven hyperfine levels shown in Fig. IV.38. The most interesting and informative new feature is, of course, the proton hyperfine interaction. The Fermi contact interaction is expected on theoretical grounds to be similar to that in H_2^+ and the observed values (450·5 MHz for H_2 and 440·1 MHz for H_2^+) are indeed close. The orbital and spin dipolar hyperfine constants were also determined for $c\ ^3\Pi_u\ H_2$ and provide stringent tests for *ab initio* calculations of the molecular wavefunction.

At the present time only preliminary accounts of the spectroscopic work on the $d\ 3\Pi_u$ state of H_2 have been reported, and we will not go into details here since it is clear that much further work is in progress in different laboratories. The general pattern of the rotational levels is presumably similar to that outlined for the $c\ ^3\Pi_u$ state. Miller and Freund (1972) have reported the detection of $M_N \rightarrow M_N + 1$ transitions in the $N = 1$ level (ortho H_2) and give a provisional analysis. They report the observation of the fine and hyperfine structure in four different vibrational levels and promise a detailed analysis in due course. Jost, Marechal and Lombardi (1972) and Cahill and Healy (1970) have made similar observations, but again a thorough analysis is awaited.

We come now to the $a\ ^3\Pi$ state of CO which has been studied extensively by Klemperer and his colleagues. The $a\ ^3\Pi$ state of CO lies about 49,000 cm^{-1} above the ground state and excitation is accomplished by electron bombardment of a carbon monoxide molecular beam, as described in Chapter II. The spectroscopic experiments are similar to the magnetic resonance experiments of Lichten, except that a static electric field replaces the magnetic field, and electric dipole transitions are studied, rather than magnetic. The deflection and focussing fields are also electrostatic, as was described in Chapter II.

$^3\Pi$ CO conforms fairly well to Hund's case (a) coupling, at least in the lower rotational levels. In case (a) coupling the electronic orbital and spin angular momenta are quantized with respect to the internuclear axis, with axial components Λ and Σ respectively, and the total component of electronic angular momentum, Ω, is well defined. Since $\Lambda = \pm 1$ and $\Sigma = \pm 1, 0$ we see that there are three fine structure states, with $|\Omega| = 0, 1$ or 2, and in $^3\Pi$ CO these are split apart by the spin-orbit coupling, the $^3\Pi_0$ state being the lowest in energy. The lower rotational levels of the $^3\Pi_0$, $^3\Pi_1$ and $^3\Pi_2$ fine structure states are illustrated in Fig. IV.39. The Λ-doubling in the $\Omega = 0$ component is large and essentially independent of J; in the $^3\Pi_1$ and $^3\Pi_2$ states the Λ-doubling (which is exaggerated in Fig. IV.39) increases with increasing J. In the dominant isotopic species, $^{12}C^{16}O$, there is no nuclear hyperfine structure but $^{13}C^{16}O$, which has $I = \frac{1}{2}$ for ^{13}C, has also been studied and will be described later.

Freund and Klemperer (1965) and Stern, Gammon, Lesk, Freund and Klemperer (1970) have carried out measurements in the $|\Omega| = 2, J = 2$ to 7 and $|\Omega| = 1, J = 1$ and 2 rotational levels of the first four vibrational levels, $v = 0$ to 3. The transitions studied are between Λ-doublet components of a given J level, either in zero field, or in the presence of a static electric field; the

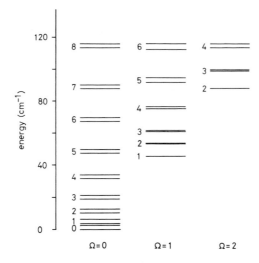

Fig. IV.39. Lower rotational levels and fine structure splitting for a $^3\Pi$ state.

transition frequencies range from 6 to 1150 MHz. Figure IV.40(a) illustrates the Stark effect in the $|\Omega| = 2$, $J = 2$ state and the radiofrequency transitions ($\Delta M_J = \pm 1$, 0) detected, whilst Fig. IV.40(b) shows a recording of the spectral lines arising from the $|M_J| = J$, $\Delta M_J = \Delta J = 0$ transition marked with an asterisk in Fig. IV.40(a). These lines, which arise from different vibrational states, form a regular series from $v = 0$ to $v = 3$, but the $v = 4$

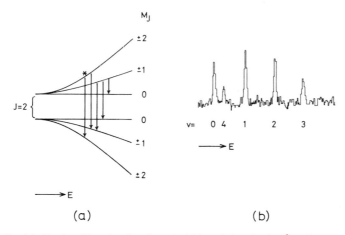

Fig. IV.40. (a) Stark effect in the $J = 2$, $|\Omega| = 2$ level of $a\,^3\Pi$ CO. (b) Electric resonance lines arising from the transition marked with an asterisk.

line is clearly anomalous, for reasons which we discuss a little later. The analysis of the data yields accurate values for the Λ-doubling intervals, and from the Stark effect the electric dipole moment in the $a\,{}^3\Pi$ state is determined. The dipole moment exhibits a measurable dependence on both the vibrational (v) and rotational (J) quantum numbers, but the variation is unusually small and a completely satisfactory quantitative explanation has yet to be provided. The actual magnitude of the dipole moment (e.g. 1·37485 D in $v = 0$, $|\Omega| = 2$, $J = 2$) provides a marked contrast with the very small dipole moment of CO in its ground electronic state (0·112 D). In terms of molecular orbital theory the ground state electron configuration of CO may be written

$$X\,{}^1\Sigma: (1\sigma)^2(2\sigma)^2(3\sigma)^2(4\sigma)^2(5\sigma)^2(1\pi)^4$$

and the very small dipole moment may then be accounted for qualitatively as arising from cancellation of the expected $C^+ - O^-$ bond polarity by the $(5\sigma)^2$ lone pair electrons on the carbon atom. The $a\,{}^3\Pi$ excited state configuration may be written

$$a\,{}^3\Pi: (1\sigma)^2(2\sigma)^2(3\sigma)^2(4\sigma)^2(5\sigma)^1(1\pi)^4(2\pi)^1$$

and the large increase in dipole moment explained by removal of one of the lone pair electrons to the antibonding 2π orbital. Calculations of the molecular wavefunction by Huo (1966) which give a good account of the ground state dipole moment predict a dipole moment for the $a\,{}^3\Pi$ state which is almost a factor of 2 too large, despite accounting for other physical constants rather well.

The anomalous position of the line arising from the $v = 4$ level, shown in Fig. IV.40(b), is due to strong perturbations between the $a\,{}^3\Pi$ and $a'\,{}^3\Sigma^+$ excited states, the latter state arising from the electron configuration

$$a'\,{}^3\Sigma^+: (1\sigma)^2(2\sigma)^2(3\sigma)^2(4\sigma)^2(5\sigma)^2(1\pi)^3(2\pi)^1.$$

The potential energy diagrams for these two states are illustrated in Fig. IV.41 and show a near-coincidence between the $v = 4$ level of the $a\,{}^3\Pi$ state and the $v = 0$ level of the $a'\,{}^3\Sigma^+$ state. Gammon, Stern and Klemperer (1971) have used the electric resonance method to study the $J = 2$ to 6 rotational levels of the $v = 4$, $|\Omega| = 2$ state, determining the Λ-doubling and Stark effect parameters. The perturbation between the two electronic states increases with increasing J because the $a\,{}^3\Pi$ ($v = 4$) state has the larger rotational constant and lies somewhat below the $a'\,{}^3\Sigma^+$ state; the coincidences become closer, however, as J increases. The perturbations arise from the admixture of the electronic states brought about by spin-orbit coupling and rotational-electronic interaction, but because of the close proximity of the states involved, the admixture cannot be treated by perturbation theory and must be handled through a more precise diagonalization procedure. This system therefore provides a good opportunity for a close look at the electronic perturbations and Gammon, Stern and Klemperer have developed the theory further than is necessary in the more usual cases when the electronic states are widely separated.

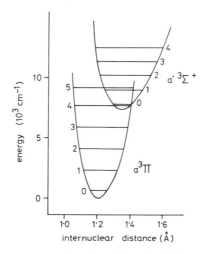

Fig. IV.41. Potential energy curves for the $a\ ^3\Pi$ and $a'\ ^3\Sigma^+$ excited electronic states of CO. Note that the $a\ ^3\Pi$ state is actually 48,687 cm^{-1} above the ground state.

The final current chapter of the $a\ ^3\Pi$ CO investigation (but probably not the last!) is the study of $^{13}C^{16}O$ by Gammon, Stern, Lesk, Wicke and Klemperer (1971). Using CO artificially enriched in ^{13}C they were able to detect and study the additional complications arising from hyperfine structure due to the nuclear spin of $\frac{1}{2}$ for ^{13}C. Zero-field and Stark effect measurements were carried out for several rotational levels in both the $^3\Pi_1$ and $^3\Pi_2$ states, with $v = 0$ to 3. From the analysis of the data Gammon et al. were able to determine the hyperfine parameters a, b, c and d, and their slight but definite dependence on v, thus providing further material for a detailed description of the molecular wavefunction.

IV.9. Concluding Remarks

The author's own feeling after reading the literature which forms the basis of this chapter is one of some awe at developments in molecular spectroscopy over the past two or three years. In particular the recent molecular beam work on excited electronic states, and especially the range and accuracy of the data, should startle those who are versed in more traditional techniques. Those whose interest lies in the calculation of molecular wave functions will realize that even for diatomic molecules the experimentalist is not yet redundant. Perhaps the most important aspect, however, is that the high resolution studies of excited electronic states in different vibrational levels are bound to lead to deeper understanding of the relations between vibrational and electronic motion.

It is difficult to predict what further developments in the field of diatomic species might occur in the near future. The molecular beam techniques are likely to find increasing applications, particularly in the study of excited electronic states where the detection techniques are so sensitive. The laser electron resonance method must also have a bright future, especially as laser technology develops. Two areas which are certainly ripe for development are the study of transition metal compounds (where beam methods could well become important) and molecular ions. Ions present a particularly difficult challenge to the experimental spectroscopist, which is not likely to be met successfully by existing conventional techniques. High resolution spectroscopy can surely provide new methods of studying energy transfer problems, particularly in crossed beam experiments. Finally one should keep a close watch on the astronomers, who are presenting spectroscopists with new problems almost daily.

V. Studies of Triatomic Species

V.1. Introduction

Only a few triatomic radicals have, as yet, been detected and studied by high resolution techniques, but enough has been done to establish the main principles and to demonstrate the major features of interest. Of the molecules which have been investigated, a very clear distinction can be drawn between linear and bent triatomic species. Open-shell linear triatomic species in orbitally *non-degenerate* electronic states are very similar to diatomic species in non-degenerate electronic states, at least so far as their rotational levels are concerned. There are differences, of course, but since no such triatomic species has yet been investigated by microwave methods, the subject rests at present. Linear triatomic species in *degenerate* electronic states are quite a different matter. NCO, for example, has a $^2\Pi$ ground electronic state and might be expected to resemble NO, except that the bending vibration in NCO exerts a profound effect on the electronic properties, with complications which introduce many fascinating aspects not encountered in diatomic molecules. Fortunately NCO has been studied in considerable detail by electron resonance and microwave methods, and a substantial part of this chapter will be devoted to a discussion of its rotational levels and spectra. NCS is the only other such species to have been investigated, and only a preliminary account of its electron resonance spectrum has as yet appeared. It is, however, similar in all major respects to NCO.

Bent triatomic species are slightly more numerous and sufficient work has now been performed to justify a fairly thorough discussion. Bent triatomic radicals with spin singlet ground states are, of course, entirely analogous in their microwave properties to stable molecules like H_2O. Those which have been studied, for example, CF_2, SiF_2 and HNO, would certainly be classified as free radicals in the chemical (kinetic) sense, but our discussion of them will be relatively brief since from a spectroscopic viewpoint they are typical asymmetric top molecules, the principal features of which are very thoroughly described in a number of other books. This is not to deny the considerable interest and importance of these species from the viewpoint of free radical chemistry. Bent triatomic species with electrons of unpaired spin are of more spectroscopic interest, however, because of the consequent magnetic interactions which provide important structural information. Up to the present time the species studied by high resolution methods all have $S = \frac{1}{2}$, and although molecules with more than one unpaired electron would be of immense interest, none has been detected by microwave techniques as yet and we therefore have no real excuse to launch into a discussion of their likely spectroscopic properties. The main examples of doublet state species are NO_2, ClO_2 and HCO. The first two are, of course, stable gases but although perhaps not free radicals in the chemical sense, they do satisfy our spectroscopic criteria and are worthy of detailed discussion. HCO satisfies all possible criteria for discussion here (including the personal interest of the author!) and therefore occupies an important part of this chapter. One aspect of the bent triatomic species with $S = \frac{1}{2}$ which will become apparent as we progress is that they are rather similar to diatomic species in $^2\Sigma$ states so far as their rotational levels and microwave properties are concerned. Molecular vibrations are of much less interest, simply because electronic spatial degeneracy cannot, by symmetry, exist for a bent triatomic molecule. Hence the strong vibrational-electronic coupling which, as we shall show, dominates the spectroscopic properties of NCO, is unimportant in HCO.

V.2. Nuclear Vibrations in a Triatomic Molecule

It will be apparent from the previous section that molecular vibrations are going to play an important part in our discussion of linear triatomic molecules. Consequently as a necessary preamble to that discussion, we first outline in a very simple manner the main aspects of vibrational motion in these species. At the level we choose for the discussion, a diagram is perhaps more helpful than the equations of motion, although it is really necessary to solve the mathematics in order to interpret the diagram correctly. Figure V.1 shows the *normal* vibrations of the nuclei in (a) a linear XY_2 molecule, (b) a linear XYZ molecule, and (c) a bent XY_2 molecule. The so-called normal vibrations as shown in Fig. V.1 arise from a particular choice of coordinates for the positions and momenta of the nuclei which allow the potential function to be written in its simplest form. For the linear XY_2 molecule we see that there are two stretching vibrations, symmetric and antisymmetric, and two degenerate bending vibrations. In the case of the linear XYZ molecule the correct picture of the normal vibrations actually depends critically upon the masses of the nuclei and the force constants of the bonds. It should also be noted that although the stretching vibrations for the linear XY_2 molecule are strictly defined, those in the XYZ molecule are linear combinations of the symmetric and antisymmetric stretching modes. (The normal vibrations for a bent XY_2 molecule are also included in Fig. V.1 but they are not going to enter into our later discussion to any significant extent.)

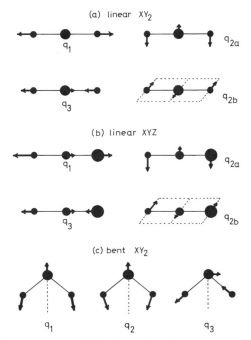

Fig. V.1. Normal vibrations in triatomic molecules.

The most important vibrations in the linear molecule are the degenerate bending vibrations. The two normal modes illustrated in Fig. V.1 are not unique solutions, however, and any linear combination of the modes shown is an acceptable solution to the vibrational wave equation. Now if we superimpose the two perpendicular vibrations q_{2a} and q_{2b} of the linear XY_2 molecule with a phase shift of $90°$, each atom will move around the linear axis in a circle, as shown in Fig. V.2, and this motion constitutes vibrational angular momentum. If the sign of the superimposition is reversed, the sense of the angular momentum is opposite to that shown in Fig. V.2.

The bending vibrational levels are characterized by values of the vibrational quantum number $v_2 = 0, 1, 2, 3$, etc. The vibrational angular momentum operator is denoted **G** and the component of vibrational angular momentum about the molecular axis is denoted l, the permissible values of l being

$$l = v_2, v_2 - 2, v_2 - 4, \ldots, -v_2. \tag{V.1}$$

It may be helpful for the reader to regard the angular momentum operator **G** for the nuclei as being analogous to **L** for the electrons, and the quantum number l (which corresponds to G_z where z is the molecular axis) as being analogous to Λ.

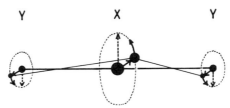

Fig. V.2. Vibrational angular momentum in a linear XY_2 triatomic molecule.

V.3. The Vibronic States of NCO

In view of the vibrational angular momentum of the nuclei discussed in the previous section, it is obviously difficult to think in terms of electronic orbital angular momentum about the molecular axis; the molecular axis no longer exists in the sense that it did for a diatomic molecule. This is a rather crude way of saying that the vibrational and electronic angular momenta will be strongly coupled and the electronic degeneracy resolved as the molecule bends. These matters were first discussed in detail by Renner (1934) and the result of the coupling between vibrational and electronic motion is known as the Renner effect. The NCO radical has a $^2\Pi$ ground electronic state and therefore provides a particularly good example of the Renner effect, and its consequences for the microwave spectra.

   The three angular momenta present in a non-rotating NCO molecule are **L**, **G** and **S**, with components along the molecular axis of Λ, l and Σ. If the spin-orbit coupling of **L** with **S** is fairly strong, Hund's case (a) provides a good description of the vector coupling and consequently a basis set in which $\Lambda\,(= \pm 1)$ and $\Sigma\,(= \pm\frac{1}{2})$ are good quantum numbers can be used. We cannot sensibly separate the vibrational and electronic angular momenta, however, and talk about electronic and vibrational states separately. Instead the resulting energy levels are called "vibronic" states, and they are characterized by a quantum number K, which is defined by

$$K = |\Lambda + l|. \tag{V.2}$$

These vibronic states are labelled according to the value of K as follows:

$$K = 0, \quad 1, \quad 2, \quad 3, \quad \text{etc.}$$

$$\text{vibronic state} \quad \Sigma \quad \Pi \quad \Delta \quad \Phi \quad \text{etc.}$$

(V.3)

Consequently we can commence our description of NCO by listing the vibronic states as shown on the left-hand side of Fig. V.3. At this stage of the

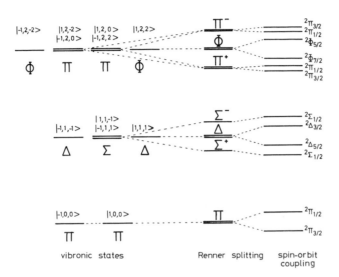

Fig. V.3. Lower vibronic states of NCO showing the vibrational splitting, Renner splitting and spin-orbit splitting.

discussion the energies of the states depend only upon the vibrational frequency v_2,

$$E(q_2) = (v_2 + 1)hv_2$$

(V.4)

and each state may be labelled according to the values of Λ, v_2 and l, i.e. $|\Lambda, v_2, l\rangle$.

Now Renner showed that the result of the vibrational-electronic coupling is to mix states which differ in Λ value by ± 2 (i.e. to mix different *electronic* states) and which differ in v_2 value by 0 or ± 2. So far as our energy level diagram is concerned, this Renner coupling leads to splitting of the different vibronic states as shown in the centre of Fig. V.3. It also leads to separation of the Σ states which were degenerate in our first description, the wave functions now corresponding to symmetric and antisymmetric combinations of the $|\Lambda, v_2, l\rangle$ labels with which we started; hence the superscripts $(+)$ and $(-)$

which appear in the labels for the vibronic states. Similar remarks apply to the Π states in the $v_2 = 2$ manifold.

Finally we must include the effects of the electron spin $S = \frac{1}{2}$. This introduces a further two-fold degeneracy into each vibronic state, but this degeneracy is resolved by the spin-orbit coupling in the same way as for a diatomic $^2\Pi$ state. Consequently the Π, Δ, Φ, etc. states are split by the spin-orbit coupling as shown on the right-hand side of Fig. V.3. The spin-orbit coupling also mixes the Σ^+ and Σ^- states and increases their separation slightly. (Consequently the $(+)$ and $(-)$ descriptions are no longer precise, although it is still convenient to use them.) The subscript in the final vibronic state label denotes the value of P, defined by

$$P = |\Lambda + l + \Sigma|, \tag{V.5}$$

whilst the pre-superscript gives the spin multiplicity $(2S + 1)$. Thus the state labels correspond to $^{2S+1}K_p$.

It should be appreciated that in constructing Fig. V.3 we have chosen to deal with the vibrational splitting first, the Renner splitting second, and the spin-orbit coupling third. We could have proceeded in the reverse order and in that case the correlation between the final vibronic states and the fine-structure states ($^2\Pi_{\frac{1}{2}}$ and $^2\Pi_{\frac{3}{2}}$) arising from consideration of *electronic* properties only would have been clearer. In ascending order of energy, the vibronic states $^2\Pi_{\frac{3}{2}}$, $^2\Delta_{\frac{5}{2}}$, $^2\Phi_{\frac{7}{2}}$, etc. in the $v_2 = 0, 1, 2$, etc. levels respectively correlate with the $^2\Pi_{\frac{3}{2}}$ *electronic* state. Similarly the $^2\Pi_{\frac{1}{2}}$, $^2\Delta_{\frac{3}{2}}$, $^2\Phi_{\frac{5}{2}}$, etc. correlate with the $^2\Pi_{\frac{1}{2}}$ *electronic* state, whilst all other states correlate with mixtures of the fine structure states. These correlations are useful when we come to relate the magnetic properties of the vibronic states to those of the electronic states of diatomic molecules. It must be emphasized, however, that the correlations are not, in general, precise.

The relative energies of the vibronic states shown in Fig. V.3 have been established by Dixon (1960) in a study of the near-ultraviolet absorption spectrum obtained by flash photolysis experiments. Dixon's measurements have recently been refined in a similar study by Bolman, Brown, Carrington and Ramsay (1973). The lowest excited electronic state in NCO is the $A\,^2\Sigma^+$ state, lying about 23,000 cm^{-1} above the ground $^2\Pi$ state, and this state does not, of course, exhibit a Renner effect since it does not possess electronic orbital degeneracy. The vibrational band structure of the electronic absorption establishes the positions of the vibronic states, and hence provides values of the relevant molecular constants. In the case of the energy level diagram shown in Fig. V.3, the main molecular constants are the spin-orbit or fine structure constant A, the vibrational frequency v_2, and a constant ε, called the Renner parameter, which is a measure of the strength of the electronic-vibrational coupling and which determines the magnitudes of the Renner splittings. The rotational structure is also beautifully resolved in the ultra-violet spectrum and its analysis yields values of the rotational constants in the different vibronic states.

Dixon's work thus provided an excellent foundation for the subsequent electron resonance and microwave investigations which we describe in the following section.

V.4. Electron Resonance and Microwave Studies of NCO and NCS

The first high resolution studies of NCO were carried out by Carrington, Fabris and Lucas (1968, 1969) who employed a continuous flow system in which fluorine atoms were mixed with the parent acid (HNCO) inside the resonance cavity of an X-band electron resonance spectrometer. They detected spectra arising from three of the vibronic states shown in Fig. V.3, namely, the ground $^2\Pi_{\frac{3}{2}}$ ($v_2 = 0$) state, the $^2\Delta_{\frac{5}{2}}$ ($v_2 = 1$) excited state and the $^2\Phi_{\frac{7}{2}}$ ($v_2 = 2$) excited state. In each case the spectrum observed arose from molecules in the lowest rotational level of the vibronic state in question; the three spectra are illustrated in Fig. V.4. The ground state spectrum is quite strong and a signal to noise ratio of 200 to 1 is readily achieved. However the $^2\Delta_{\frac{5}{2}}$ spectrum is weaker, and the $^2\Phi_{\frac{7}{2}}$ spectrum weaker still, because these states lie well above the ground state and their thermal population is therefore low. Indeed the $^2\Phi_{\frac{7}{2}}$ spectrum was only observed after signal to noise enhancement using computer averaging techniques.

In many ways the analysis of these spectra follows along lines similar to those for $^2\Pi$ diatomic molecules. Taking the ground vibronic state spectrum, for example, in the lowest rotational level J is equal to $\frac{3}{2}$ and the magnetic field removes the $2J + 1$ degeneracy. There are therefore three $\Delta M_J = \pm 1$ transitions which give rise to the three separated groups. The group separation depends upon the second-order Zeeman effect discussed previously, and is inversely proportional to the size of the rotational constant. Each group exhibits a further triplet splitting arising from hyperfine interaction with ^{14}N. For the $^2\Delta_{\frac{5}{2}}$ spectrum the J value in the lowest rotational level is $\frac{5}{2}$ and, as a result, there are five $\Delta M_J = \pm 1$ transitions giving rise to five groups. Similarly for the $^2\Phi_{\frac{7}{2}}$, $J = \frac{7}{2}$ spectrum we have seven groups.

The main feature of interest in each spectrum is its g value, listed below. To a first approximation the theoretical g value is given by the simple expression,

$$g_J = \frac{(g_L\Lambda + g_e\Sigma)(\Lambda + \Sigma + l)}{J(J + 1)} \tag{V.6}$$

and if we put $g_L = 1$ and $g_e = 2.0023$ we obtain the following first-order g values

$$^2\Pi_{\frac{3}{2}}, J = \tfrac{3}{2}: g = 0.8005, \qquad ^2\Delta_{\frac{5}{2}}, J = \tfrac{5}{2}: g = 0.5718,$$
$$g(\text{exp}) = 0.7909, \qquad\qquad g(\text{exp}) = 0.5638,$$

$$^2\Phi_{\frac{7}{2}}, J = \tfrac{7}{2}: g = 0.4447.$$
$$g(\text{exp}) = 0.4356.$$

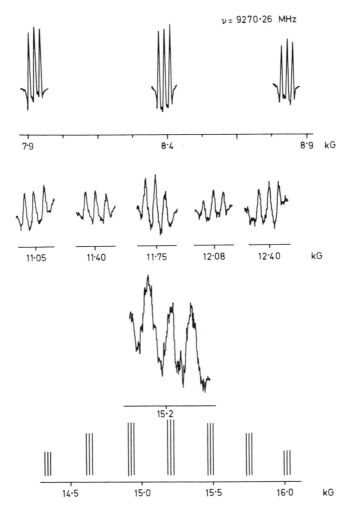

Fig. V.4. Gas phase electron resonance spectra of NCO. *Top:* $^2\Pi_{\frac{3}{2}}$ ($v_2 = 0$), $J = \tfrac{3}{2}$. *Centre:* $^2\Delta_{\frac{5}{2}}$ ($v_2 = 1$), $J = \tfrac{5}{2}$. *Bottom:* $^2\Phi_{\frac{7}{2}}$ ($v_2 = 2$), $J = \tfrac{7}{2}$.

Although at first sight the predicted first-order and measured g values might seem to be quite close, the differences are, in fact, substantial. With a satisfactory theory one might expect agreement between the calculated and observed g values to be within ± 0.0001, so we have some way to go. There are, indeed, a number of effects not taken into account by the simple formula, which we now examine one at a time.

(i) There are small relativistic corrections to g_L and g_e due to the relatively high velocity of the electrons. After inclusion in the theory the calculated g values for the three vibronic states are

$$0.8004, \qquad 0.5717, \qquad 0.4447.$$

(ii) The magnetic field mixes different J levels, giving rise to the second-order separation of the $\Delta M_J = \pm 1$ transitions discussed earlier. This same rotational level mixing also affects the g values; inclusion in the theory leads to predicted g values

$$0.7887, \qquad 0.5666, \qquad 0.4420.$$

(iii) The rotation of the molecule leads to mixing of the fine structure states; the extent of this mixing is proportional to the rotational constant B, and inversely proportional to the fine structure constant A. After inclusion of this effect the theoretical g values are

$$0.7952, \qquad 0.5713, \qquad 0.4456.$$

The corrections up to this point also arise in the theory of diatomic molecules. Close examination of the theory of the Renner effect by Carrington, Fabris, Howard and Lucas (1971) revealed that the g values are also affected by mixing of the $^2\Pi$ vibronic states with $v_2 = 0$ and 2; this effect does not, of course, arise in diatomic molecules but in NCO its inclusion leads to predicted g values of

$$0.7932, \qquad 0.5668, \qquad 0.4388.$$

If we now compare these predicted values with the measured ones, the differences for the three vibronic states studied are

$$-0.0023, \qquad -0.0030, \qquad -0.0032.$$

These discrepancies are still large, particularly when converted to magnetic field positions. We note that the discrepancy increases with increasing vibrational quantum number and Carrington et al. (1971) therefore extended the Renner theory to higher order and showed that, indeed, a further correction to the Zeeman effect should arise due to partial quenching of the orbital angular momentum as the molecule bends. According to their theory the discrepancies given above should be equal to $(v_2 + 1) \Delta g_L/J + 1$, where Δg_L is a small negative constant determined by the strength of the vibrational-electronic coupling. Equating this expression with the g value discrepancies, the Δg_L value for each vibronic state is then found to be

$$^2\Pi_{\frac{3}{2}}, \Delta g_L = -0.0058, \qquad ^2\Delta_{\frac{5}{2}}, \Delta g_L = -0.0053, \qquad ^2\Phi_{\frac{7}{2}}, \Delta g_L = -0.0048.$$

Carrington *et al.* were disappointed to find that Δg_L was apparently not constant, as their theory predicted, but still seemed to exhibit an apparent variation from one vibronic state to the next. They pointed out that this might be due to anharmonic effects not included in the theory, but also noted that the apparent variation in Δg_L would disappear if the value of the Renner parameter ε was somewhat reduced in magnitude. This observation led to the re-measurement of the ultraviolet absorption spectrum by Bolman *et al.* mentioned earlier, who confirmed that there were minor errors in Dixon's earlier work and that a better value of ε is -0.146. Inclusion of this value in the theory yields the following value of Δg_L for the three vibronic states:

$$\Delta g_L = -0.0065 \pm 0.0002.$$

Thus within the limits of experimental error Δg_L is, indeed, found to be constant; the electron resonance and ultraviolet studies are now consistent with each other and the present theory of the Renner coupling is evidently fairly satisfactory. There are further tests which should be made, however, which we will outline a little later.

The triplet hyperfine splitting from ^{14}N arises in much the same way as does the ^{35}Cl quartet splitting in ClO (see section IV.6.3). The measurements yield a value for $a + \frac{1}{2}(b + c)$ and also the ^{14}N electric quadrupole coupling constant. Within the limits imposed by their line width, Carrington *et al.* were unable to detect any changes in the values of these two constants between the three vibronic states studied. In further experiments they studied the Stark splittings produced by static electric fields and determined the electric dipole moment to be 0.742 D; again no significant vibrational dependence was detected.

Following the detection of NCO by electron resonance methods, Saito and Amano (1970) and Amano and Hirota (1972) used the same chemical approach to generate a sufficient concentration of NCO for pure microwave studies. Saito and Amano studied the $J = \frac{3}{2} \to \frac{5}{2}$ (~ 58.1 GHz) and $\frac{5}{2} \to \frac{7}{2}$ (~ 81.4 GHz) rotational transitions in the ground $^2\Pi_{\frac{3}{2}}$ state, and the $J = \frac{3}{2} \to \frac{5}{2}$ (~ 58.6 GHz) rotational transition in the $^2\Pi_{\frac{1}{2}}$ ($v_2 = 0$) state. The $^2\Pi_{\frac{1}{2}}$ spectrum shows splitting due to Λ-doubling from which the α_Π parameter was determined. Saito detected $\Delta F = 0$ and ± 1 transitions in the $^2\Pi_{\frac{3}{2}}$ state, and $\Delta F = \pm 1$ transitions in the $^2\Pi_{\frac{1}{2}}$ state; consequently he was able to effect a partial separation of the magnetic hyperfine constants, determining unique values of a, $(b + c)$ and d, in addition to the quadrupole coupling constant. Amano and Hirota have recently detected rotational transitions in the $v_2 = 1$, $^2\Delta_{\frac{5}{2}}$ and $^2\Delta_{\frac{3}{2}}$ states, and have also refined Saito's earlier measurements in the $v_2 = 0$ states. They find that there is, in fact, a measurable vibrational dependence of the hyperfine constants and give the following values:

$^2\Delta\ (v_2 = 1)$ $a = 60\cdot47$ MHz, $b + c = -17\cdot33,$ $eqQ = -2\cdot11.$

$^2\Pi\ (v_2 = 0)$ $a = 64\cdot31$ MHz, $b + c = -21\cdot20,$ $eqQ = -1\cdot85.$

It will be interesting to see if data for higher vibrational states can be obtained, particularly with regard to the hyperfine constants which are such delicate probes of the electronic wave function. Electron resonance studies of the $v_2 = 1$, $^2\Sigma$ states are also desirable because these states are more sensitive to the Renner effect than those investigated so far, and would provide stringent tests of the Renner effect theory. They are expected to be strongly magnetic and therefore amenable to resonance studies; conversely the $^2\Pi_{\frac{1}{2}}\ (v_2 = 0)$, $^2\Delta_{\frac{3}{2}}\ (v_2 = 1)$ and $^2\Phi_{\frac{5}{2}}\ (v_2 = 2)$ states correlate strongly with the $^2\Pi_{\frac{1}{2}}$ *electronic* state and are therefore almost non-magnetic.

The only other linear triatomic molecule to have been studied by high resolution spectroscopy is NCS. Carrington, Fabris and Lucas (1969) generated a detectable concentration through the reaction of F atoms with HNCS and made preliminary measurements of the spectra arising from the ground $^2\Pi_{\frac{3}{2}}\ (v_2 = 0)$ and excited $^2\Delta_{\frac{5}{2}}\ (v_2 = 1)$ states. The spectra are similar to those of NCO but require much more detailed attention than has yet been given; the experiments are difficult because production of the radical is accompanied by the deposition of elementary sulphur in the resonance cavity.

The electron resonance and microwave studies of NCO provide excellent examples of the complementary nature of these techniques, and also the depth in understanding of molecular properties which spectroscopy can provide. The Renner effect is important because it is pertinent to fundamental questions about the inter-dependence of electronic and nuclear motions in molecules; it is to linear molecules what the Jahn-Teller effect is to non-linear molecules. It is a curious fact that the Jahn-Teller effect exerts a fascinating hold over theorists, whilst being almost completely elusive to the experimental spectroscopist. In contrast, the Renner effect has been studied very thoroughly in the laboratory, but the problems raised have yet to be tackled seriously by those interested in the calculation of molecular wavefunctions.

V.5. Microwave Spectra of HNO, CF_2 and SiF_2

We mentioned in the introductory section of this chapter that three closed shell bent triatomic radicals have been detected and studied by microwave spectroscopy; these are HNO, CF_2 and SiF_2. They are typical asymmetric top molecules and our description will be fairly brief. The open-shell radicals to be described later are more complicated and their microwave spectra are consequently more informative. However it is advantageous to discuss first the closed-shell systems since they exhibit features which also arise in the open-shell radicals. Thus, so far as the rotational levels are concerned, HNO is similar to HCO, but without the additional complications which arise because of the presence of unpaired electron spin in HCO.

HNO and DNO have been detected recently by Saito and Takagi (1972), DNO being the more extensively studied at present. The DNO radical is produced in a continuous flow system by the reaction of deuterium atoms with NO; its lifetime is found to range from 1 to 40 seconds depending upon the pressure and other chemical species present. The mixing is arranged to occur inside a parallel plate microwave cell of the type described in Chapter II, and the microwave spectrometer is of the conventional 100 kHz Stark modulation type. The transitions detected are best discussed in terms of the rotational level diagram shown in Fig. V.5. DNO is a near-prolate top ($A = 315450 \cdot 3$ MHz, $B = 38731 \cdot 5$, $C = 34354 \cdot 0$) and, as we discussed previously in section III.5.5, the rotational levels can be arranged in stacks corresponding to $K = 0, \pm 1, \pm 2$, etc., each stack consisting of levels characterized by values of N. For $K > 0$ each N value characterizes a pair of closely spaced levels, called K-doublets.

The transitions in DNO detected by Saito and Takagi (1972) are of three types, as listed below:

a-type	$0_{00} \rightarrow 1_{01}$	$\sim 73 \cdot 1$ GHz
b-type	$3_{03} \rightarrow 2_{12}$	$\sim 53 \cdot 3$
	$3_{13} \rightarrow 4_{04}$	$\sim 25 \cdot 8$
K-doubling	$4_{14} \rightarrow 4_{13}$	$\sim 43 \cdot 7$
	$5_{15} \rightarrow 5_{14}$	$\sim 65 \cdot 6$

These transitions are shown in Fig. V.5 and their nomenclature may be understood in terms of the axis system for DNO illustrated in Fig. V.6(a). The bond angle in DNO is close to $109°$ and the a and b inertial axes are as shown, with the c axis perpendicular to the molecular plane. a-type transitions are induced

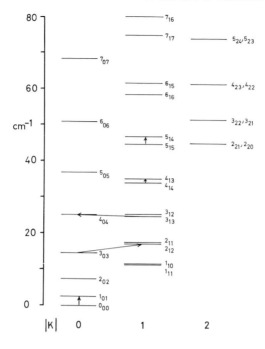

Fig. V.5. Lower rotational levels of DNO and observed transitions.

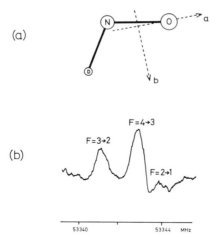

Fig. V.6. (a) Inertial axes of DNO. (b) ^{14}N quadrupole hyperfine structure of the $0_{00} \rightarrow 1_{01}$ transition in DNO.

by coupling between the electric vector of the microwave field and the component of the molecular electric dipole moment along the a inertial axis. Similarly the b-type transitions involve the dipole moment component along the b axis. There are no c-type transitions in DNO because, of course, the component of the dipole moment perpendicular to the plane of the molecule is zero. Note, incidentally, that the K-doubling transitions are actually a-type. Takagi and Saito (1972) measured the Stark effect on the a- and b-type transitions, determining the respective dipole moment components to be $\mu_a = 1 \cdot 18$ D, $\mu_b = 1 \cdot 22$ D. The resultant total dipole moment is $1 \cdot 70$ D, its direction very nearly corresponding to the bisector of the angle between the a and b axes.

The rotational levels shown in Fig. V.5 do not include any nuclear hyperfine effects. There are, of course, no magnetic interactions involving electron spin but there is an electric quadrupole interaction with ^{14}N and the transitions studied by Takagi and Saito all exhibited quadrupole hyperfine structure except for the $0_{00} \rightarrow 1_{01}$ transition. Figure V.6(b) shows the hyperfine structure of the $3_{03} \rightarrow 2_{12}$ transition, and from the measurement of these components the magnitude of the quadrupole coupling can be determined. Now the quadrupole coupling is actually described by a tensor and is specified in terms of components in the a, b and c directions, which were determined to be

$$(eqQ)_{aa} = 1 \cdot 03 \text{ MHz}, \qquad (eqQ)_{bb} = -6 \cdot 13, \qquad (eqQ)_{cc} = 5 \cdot 1.$$

The precise definition of these components is fairly complicated and the interested reader should consult the original paper. (In this connection it should be noted that the so-called "quadrupole coupling constant" in a diatomic molecule is actually the component of the quadrupole tensor along the internuclear axis. The components perpendicular to the internuclear axis are not necessarily zero, neither are they necessarily equal except in a pure Σ state.)

Takagi and Saito have also made measurements on the HNO radical and determined values of the same molecular constants as have been described for DNO. There is a chance that HNO might be abundant in interstellar gas clouds and the measurement of the lowest rotational transition frequency ($81477 \cdot 49 \pm 0 \cdot 1$ MHz) is clearly an important first step in the possible identification of this species by astronomers.

The microwave spectra of CF_2 and SiF_2 have been reported but so far the detailed analyses have not been published. CF_2 is produced by passing C_2F_3Cl, CF_4 or $(CF_3)_2CO$ through a radiofrequency discharge. Microwave transitions ranging from $21 \cdot 5$ to $84 \cdot 8$ GHz have been detected by Powell and Lide (1966) from which the following rotational constants have been determined:

$$A = 88351 \cdot 65 \text{ MHz}, \qquad D = 12507 \cdot 55, \qquad C = 10931 \cdot 85.$$

Hence CF_2 is also a near-prolate top and the rotational level diagram is similar to that shown for DNO, except that the rotational spacings are smaller. The a inertial axis lies parallel to a line through the two fluorine atoms, the FCF bond angle being 104·9°, and since the electric dipole moment component in this direction must be zero by symmetry, it follows that only b-type transitions are possible. The dipole moment lies along the two-fold symmetry axis of the molecule (which is also the b inertial axis) and preliminary measurements indicate a dipole moment of about 0·46 D.

The SiF_2 radical can be prepared by passing SiF_4 over heated silicon. It is a relatively long-lived species and a preliminary report of its microwave spectrum in the range 8–30 GHz has been given by Rao, Curl, Timms and Margrave (1965). The rotational constants are

$$A = 30601{\cdot}52 \text{ MHz}, \qquad B = 8823{\cdot}91, \qquad C = 6830{\cdot}55,$$

so that again it is a near-prolate top, with an FSiF bond angle of 100° 59′. The microwave transitions are again necessarily b-type transitions and the electric dipole moment (lying in the direction of the b axis) is 1·23 D.

The CF_2 and SiF_2 microwave spectra do not exhibit resolved nuclear hyperfine effects, there being no nucleus with a quadrupole moment and, of course, no unpaired electron spin. Current work on these molecules is directed towards an understanding of the centrifugal distortion.

V.6. Electron Resonance and Microwave Studies of HCO and DCO

The visible and near-ultraviolet spectrum of HCO has been measured and analysed by Herzberg and Ramsay (1955) and Johns, Priddle and Ramsay (1963). The rotational constants were found to be

$$A = 670480 \text{ MHz}, \qquad B = 44801, \qquad C = 41995,$$

the bond angle being 119° 30′ and the CH and CO bond lengths being 1·08 Å and 1·198 Å respectively in the ground vibrational state. The radical was produced by flash photolysis of acetaldehyde. Recently Bowater, Brown and Carrington (1971, 1973) found that the reaction of fluorine atoms with formaldehyde gives a sufficient concentration of HCO in a continuous flow system for it to be detected by electron resonance. Subsequently Saito (1973) has used the same method of generation to measure part of the pure microwave spectrum. Comparison of the two techniques as applied to HCO is of considerable interest, and points the way to possible future developments in the study of non-linear open-shell radicals.

The rotational constants for HCO are similar to those of HNO so that the pattern of rotational levels is similar to that already described in Fig. V.5. We must, however, now consider the additional complications introduced by the electron spin ($S = \frac{1}{2}$) in HCO and to this end should write the appropriate effective Hamiltonian. It is as follows:

$$\mathcal{H}_{\text{eff}} = A_0 N_a^2 + B_0 N_b^2 + C_0 N_c^2$$
$$+ \{\varepsilon_{aa}(N_a S_a) + \varepsilon_{bb}(N_b S_b) + \varepsilon_{cc}(N_c S_c) + \tfrac{1}{2}(\varepsilon_{ba} + \varepsilon_{ab})(N_a S_b + S_a N_b)\}$$
$$+ a_C \mathbf{S.I} + \mathbf{S.T.I}. \tag{V.7}$$

The first three terms represent the normal asymmetric top Hamiltonian for the nuclear rotation. The fourth term describes the electron spin-nuclear rotation coupling and, as we discussed earlier in section IV.5 for $^2\Sigma$ diatomic radicals, arises in part from the direct interaction between the spin and nuclear rotational magnetic moments, and in part from spin-orbit mixing of excited electronic states with the ground state. We shall say more about this term in due course. The fifth term in equation (V.7) is the Fermi contact interaction between the electron and proton spins, whilst the last term describes the dipolar coupling of the spins. We will discuss the expansion of the dipolar coupling in terms of components along the a, b, c inertial axes later.

The net result of the spin-rotation interaction is to split each rotational level into a doublet, the resulting levels being characterized by values of $J = N \pm \frac{1}{2}$. This spin doubling is too small to be resolved in the optical spectrum, but is certainly expected to be large compared with the resolution obtainable in a microwave experiment. Each J level is further split into a doublet by the hyperfine terms, the final levels being characterized by values of the grand total angular momentum quantum number $F = J \pm \frac{1}{2}$.

Now for reasons which should become clear later, Bowater, Brown and Carrington looked for a rotational transition whose frequency was likely to be close to the operating frequency of their X-band electron resonance spectrometer. From the optical values of the rotational constants, the K-doubling transition $2_{12} \rightarrow 2_{11}$ appeared to be suitable, the transition frequency being predicted to be $3(B - C) = 8418$ MHz. This estimate did not, of course, include any consideration of the spin-rotation or nuclear hyperfine splittings, whose magnitudes were unknown. Nevertheless the estimate was good enough for the electron resonance search to be successful. Figure V.7 shows the spin-rotation and nuclear hyperfine splitting of these K-doublet components, and the four $\Delta F = 0$ transitions detected by Bowater et al. However the electron resonance experiments are performed by keeping the microwave frequency fixed and varying the magnetic field. Consequently the next step is to see how the levels shown in Fig. V.7 behave in a magnetic field.

The Hamiltonian for the Zeeman effect may be written

$$\mathscr{H}_Z = g_e\mu_B\mathbf{B}.\mathbf{S} + \mu_B\mathbf{B}.\mathbf{L} - g_I\mu_N\mathbf{B}.\mathbf{I}, \tag{V.8}$$

the three terms representing the interaction of the magnetic field with the contributions to the molecular magnetic moment arising from electron spin, electron orbital angular momentum and nuclear spin. There should also be a term involving the rotational magnetic moment but its exact form for an asymmetric top is complicated, its magnitude is small and it is by no means

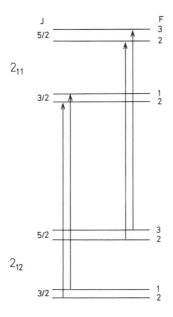

Fig. V.7. Spin-rotation and proton hyperfine splitting of the $2_{12} \rightarrow 2_{11}$ K-doubling transition in HCO.

easily measured. Since the direction of the external magnetic field defines the space-fixed z direction, each scalar product in (V.8) can be reduced to the z component. The net effect of the magnetic field is to remove the $2F + 1$ degeneracy of each of the eight levels, as illustrated in Fig. V.8. The transitions studied by Bowater *et al.* obeyed the selection rules $\Delta N = \Delta J = \Delta F = 0$, as shown in Fig. V.7, and also $\Delta M_F = \pm 1$. Now it is apparent from Fig. V.8 that the Zeeman splitting of each level is initially linear (first-order) and if this were true for all magnetic field strengths, it would clearly be possible to tune each $\Delta M_F = \pm 1$ transition into resonance by varying the magnetic field,

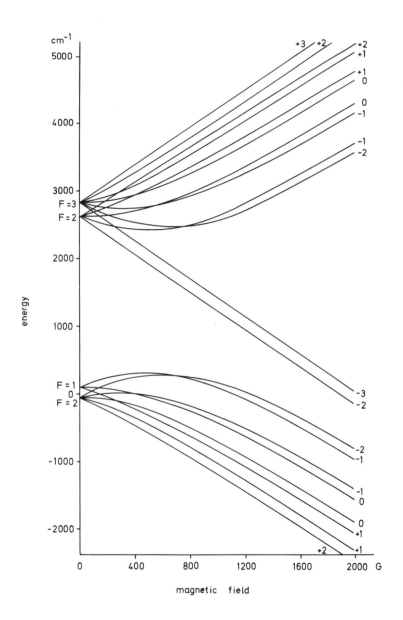

Fig. V.8. Zeeman effect in the 2_{12} or 2_{11} levels of HCO. The Zeeman components are labelled by their M_F values.

even if the spectrometer frequency was considerably different from the zero-field frequency. However Fig. V.8 shows that the Zeeman effect rapidly becomes second-order (non-linear) and the $\Delta M_F = \pm 1$ transition frequencies become less tunable (or lose their electric dipole intensity). This important result is illustrated in Figs V.9 and V.10, which show the transition frequencies as functions of applied field strength. Those transitions which retain their electric dipole intensity can be tuned only over a limited range by varying the magnetic field. A further important feature is that in the linear region the different $\Delta M_F = \pm 1$ components belonging to a particular $\Delta N = \Delta J = \Delta F = 0$ transition are not split, but as soon as we approach the non-linear region the components become separated and are therefore weaker. This second-order Zeeman splitting was observed by Bowater $et\ al.$, but is difficult to study in detail because of the low intensity of the lines.

There are, of course, good physical reasons for these features. The electron resonance spectroscopist seeks a pair of levels whose separation can be varied

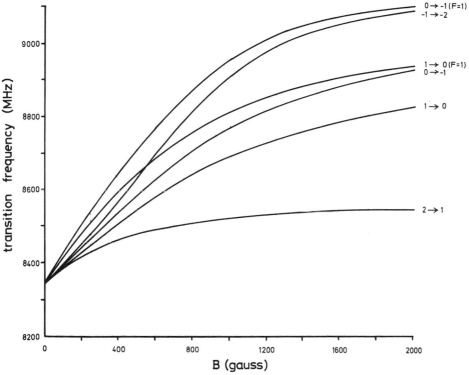

Fig. V.9. Transition frequencies of the $\Delta M_F = \pm 1$, $J = \frac{3}{2}$ K-doublet components as a function of applied magnetic field strength.

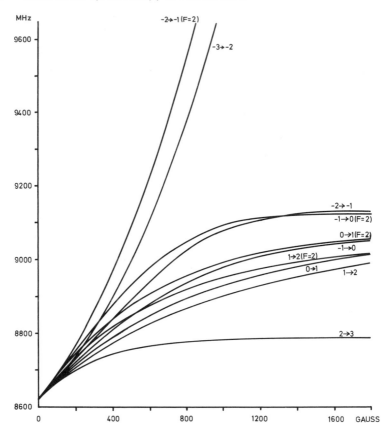

Fig. V.10. Transition frequencies of the $\Delta M_F = \pm 1$, $J = \frac{5}{2}$ K-doublet components as a function of applied magnetic field strength.

over a large range by changing the magnetic field, but which are still connected by an electric dipole matrix element. The magnetic tunability arises primarily from the electron spin, but the transition only retains its electric dipole character if the spin orientation is strongly coupled to the molecular orientation. If there is no magnetic field present, the spin is coupled to the molecular orientation through the spin-rotation interaction term, but this is relatively small in magnitude. Consequently the orienting effect of the applied field rapidly overcomes that of the spin-rotation coupling and, as we see from the right-hand side of Fig. V.8, the levels separate into two diverging groups, which correspond to $M_S = \pm \frac{1}{2}$. This decoupling effect represents the main limiting feature of electron resonance techniques in the study of non-linear radicals (and also linear radicals in $^2\Sigma$ electronic states). Except in a few

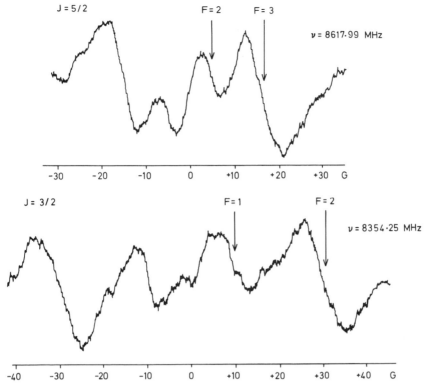

Fig. V.11. Gas phase electron resonance spectrum of HCO ($2_{12} \rightarrow 2_{11}$ transitions).

special cases, the spin-rotation coupling is weak and easily broken by the applied field. Consequently the spectrometer frequency must be fairly close to the zero-field transition frequency.

Bowater, Brown and Carrington studied each of the four transitions shown in Fig. V.7 by recording the spectra at a series of set microwave frequencies and sweeping the magnetic field. A typical spectrum is shown in Fig. V.11 and by confining their measurements to the linear Zeeman effect region, they were able to extrapolate their measurements to zero magnetic field quite accurately. The extrapolation is shown in Fig. V.12 and it yields the following zero-field transition frequencies:

$$N = 2 \quad J = \tfrac{3}{2} \quad F = 1 \quad \nu = 8347\cdot3 \text{ MHz}$$
$$2 \qquad \tfrac{3}{2} \qquad 2 \qquad 8340\cdot5$$
$$2 \qquad \tfrac{5}{2} \qquad 2 \qquad 8621\cdot0$$
$$2 \qquad \tfrac{5}{2} \qquad 3 \qquad 8625\cdot2$$

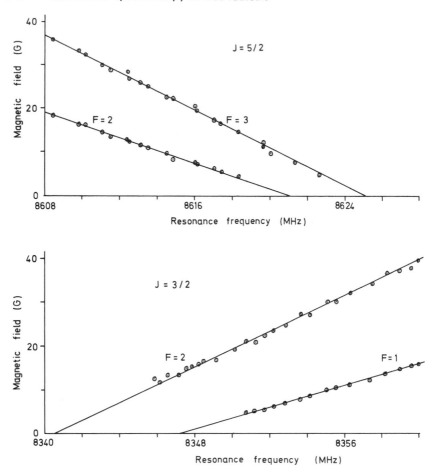

Fig. V.12. Zero-field extrapolation of the electron resonance transitions in HCO.

One might reasonably ask whether this complicated process is worthwhile in comparison with attempts to measure the zero-field transition frequencies directly. In fact the transitions have been detected subsequently with swept-frequency techniques by Levy (1972) and Radford (1972). However it is easier to detect a spectrum when the means of production of the radical and the transition frequencies have already been established. Swept-frequency experiments in the range required for HCO are not easy, particularly in comparison with swept-field methods. There are at present very few contradictions to the statement that free radical spectra are more readily first detected by resonance experiments, and subsequently studied in detail by swept-frequency methods.

Measurements of four transition frequencies take us a little way towards the goal of determining all the constants which appear in the effective Hamiltonian. Bowater, Brown and Carrington showed that the observed frequencies depend primarily on three parameters:

$$B_0 - C_0 = 2837\cdot1 \pm 2\cdot0 \text{ MHz}$$
$$\tfrac{1}{2}(\varepsilon_{bb} - \varepsilon_{cc}) = 112\cdot5 \pm 0\cdot4$$
$$T_{bb} - T_{cc} = 19\cdot2 \pm 1\cdot4$$

There is also a very slight dependence on the Fermi contact interaction, which mixes the $J = \tfrac{3}{2}$ and $\tfrac{5}{2}$ states.

More recently Saito (1973) has measured the lowest rotational transition, $0_{00} \rightarrow 1_{01}$, using a parallel plate cell and conventional microwave spectroscopy at a frequency of 86·7 GHz. The lowest rotational level (0_{00}) is not split by the spin-rotation interaction but the 1_{01} level is split, as shown in Fig. V.13(a). Saito measured three transitions, as indicated, and obtained values for three more molecular parameters:

$$B_0 + C_0 = 86718\cdot4 \pm 0\cdot2 \text{ MHz}$$
$$a_C = 372\cdot2 \pm 0\cdot4$$
$$\tfrac{1}{2}(\varepsilon_{bb} + \varepsilon_{cc}) = -95\cdot2$$

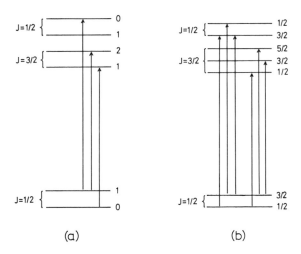

(a)   (b)

Fig. V.13. (a) Spin-rotation and proton hyperfine splitting of the lowest rotational levels in HCO, and the transitions observed by Saito (1973). (b) Spin-rotation and deuteron hyperfine splitting of the lowest rotational levels in DCO, and the transitions observed by Saito (1973).

Consequently separate values of B_0 and C_0 may be obtained by combining the electron resonance and pure microwave results ($B_0 = 44778$ MHz, $C_0 = 41940$ MHz) and these agree (within experimental error) with the optical values. It is also possible to obtain unique values of ε_{bb} and ε_{cc} which are

$$\varepsilon_{bb} = \quad +17\cdot3 \text{ MHz}$$
$$\varepsilon_{cc} = -207\cdot7 \text{ MHz}$$

The remaining diagonal component of the spin-rotation tensor (ε_{aa}) could be determined through measurement of a $\Delta J = \pm 1$ transition, either a K-doubling or a rotational ($\Delta N = \pm 1$) transition, for any rotational levels where K is greater than zero. The dipolar hyperfine tensor should be determinable from Bowater et al. and Saito's results, but the coupling is quite small and greater accuracy in the measurements is required.

It is possible to relate the gas phase studies of HCO to solid state electron spin resonance investigations. The HCO radical in a solid carbon monoxide matrix at 4·2 K has been studied by Adrian, Cochran and Bowers (1962), and in a single crystal of formic acid at 77 K by Holmberg (1969). The electron spin resonance spectrum yields values of the Fermi contact interaction, the principal components of the dipolar hyperfine tensor, and also the principal components of the g tensor. We recall from section III.2.6 that the g tensor arises from a derivation of the effective Hamiltonian in which the spin-orbit mixing of excited electronic states with the ground state is taken into account. The Zeeman interaction term is then written

$$\mathcal{H}_Z = \mu_B \mathbf{B} \cdot \mathbf{g} \cdot \mathbf{S} \tag{V.9}$$

instead of the usual sum of terms

$$\mathcal{H}_Z = g_e \mu_B \mathbf{B} \cdot \mathbf{S} + \mu_B \mathbf{B} \cdot \mathbf{L}. \tag{V.10}$$

The g tensor in equation (V.9) is therefore related to the second-order part of the spin-rotation tensor, since both involve the spin-orbit coupling. We will say more about this later, but first we outline the solid state results. Adrian, Cochran and Bowers determined the principal values of the g tensor to be

$$g_1 = 1\cdot9960, \qquad g_2 = 2\cdot0027, \qquad g_3 = 2\cdot0041,$$

but they were unable to determine the directions of the principal axes, although one axis must, by symmetry, be perpendicular to the molecular plane. Holmberg subsequently gave the principal values of the g tensor as

$$g_1 = 1\cdot9948, \qquad g_2 = 2\cdot0023, \qquad g_3 = 2\cdot0037,$$

which agree quite well with the earlier values. Again, he was unable to locate the directions of the axes, but there are good theoretical reasons for assuming that g_1 refers to the C—O bond direction, g_2 is in-plane but perpendicular to the C—O bond, and g_3 is perpendicular to the molecular plane. This assignment is illustrated in Fig. V. 14.

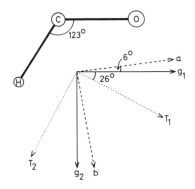

Fig. V.14. Location of the in-plane principal axes of the inertial, g and dipolar (T) tensors for HCO.

The isotropic proton hyperfine constant is given by Adrian, Cochran and Bowers as 383·9 MHz, by Holmberg as 354 MHz, whilst the gas phase value from Saito's work is 372 MHz. The differences between these values are substantial and suggest that the solid state environment has a perturbing effect on the contact interaction. The dipolar hyperfine tensor as determined by Holmberg has principal values

$$T_1 = +25 \text{ MHz}$$
$$T_2 = -8 \text{ MHz}$$
$$T_3 = -17 \text{ MHz}$$

and it seems most likely that the directions of the principal axes are as shown in Fig. V.14. We can compare these results with the gas phase measurements by an in-plane rotation of the T tensor from the (1, 2, 3) axis system to the (a, b, c) inertial axis system; the result is that the "solid state" value of $T_{bb} - T_{cc}$ is 18 MHz, which agrees well with the gas phase result (19·2 MHz). However the comparison of gas phase and solid state results is not yet sufficiently critical for one to make a reliable assessment of the effects of the solid state environment. It is very important that such an assessment be made, because many of our current views about the molecular and electronic structure of free radicals are based on solid state electron spin resonance results, with the assumption that the environmental perturbing effects are small.

As we have already mentioned, the solid state g tensor and gas phase ε tensor are closely related. If we define Δg to be equal to $g(\text{observed}) - g_e$, where g_e is the free electron value (2·0023), Curl (1965) has shown that the principal components of the g tensor are given by

$$\Delta g_a = -\frac{\varepsilon_{aa}}{2A}, \qquad \Delta g_b = -\frac{\varepsilon_{bb}}{2B}, \qquad \Delta g_c = -\frac{\varepsilon_{cc}}{2C}. \qquad \text{(V.11)}$$

These relations are only approximately true but when applied to HCO they do at least show that the signs of the principal components of the ε tensor and the corresponding Δg values are consistent. Comparison of the magnitudes is more difficult because of the uncertainty of the location of the principal axes of the g tensor. Furthermore we are neglecting first-order contributions to the spin-rotation tensor, attributing the observed constants entirely to the spin-orbit coupling of the ground and excited electronic states. Again, more work is needed before a reliable comparison of the solid state and gas phase results can be made.

DCO has now been studied at least as extensively as HCO; it is generated by the reaction of fluorine atoms with D_2CO. Electron resonance studies of the $2_{12} \rightarrow 2_{11}$ K-doubling transition have again been made, the main difference from HCO being that the K-doubling transition frequency is somewhat higher, and the spin doublet (J) components are split into triplets through hyperfine interaction with the deuteron which has $I = 1$. The following molecular constants have been determined by Bolman, Brown, Carrington and Lycett (1973):

$$
\begin{aligned}
B_0 - C_0 &= 3290\text{·}7 \text{ MHz} \\
\tfrac{1}{2}(\varepsilon_{bb} - \varepsilon_{cc}) &= \quad 89\text{·}7 \\
\varepsilon_{aa} &= 6450\text{·}0 \\
T_{bb} - T_{cc} &= \quad 3\text{·}0
\end{aligned}
$$

Similarly the $0_{00} \rightarrow 1_{01}$ rotational transition (see Fig. V.13(b)) has been studied by Saito (1973) and he gives the following values of the molecular constants:

$$
\begin{aligned}
B_0 + C_0 &= 73533\text{·}9 \text{ MHz} \\
\tfrac{1}{2}(\varepsilon_{bb} + \varepsilon_{cc}) &= \quad -84\text{·}2 \\
T_{bb} + T_{cc} &= \quad -1\text{·}4 \\
a_C &= \quad 58\text{·}7
\end{aligned}
$$

Hence by combining the two sets of data it is possible to obtain unique values for the individual constants, as follows:

$$B_0 = 38412 \cdot 3 \text{ MHz}, \qquad C_0 = 35121 \cdot 6,$$

$$\varepsilon_{bb} = 5 \cdot 45, \qquad \varepsilon_{cc} = -173 \cdot 9, \qquad \varepsilon_{aa} = 6450,$$

$$T_{aa} = 1 \cdot 40, \qquad T_{bb} = 0 \cdot 82, \qquad T_{cc} = -2 \cdot 22, \qquad a_C = 58 \cdot 7.$$

Note that T_{aa} is determined through the relation $T_{aa} + T_{bb} + T_{cc} = 0$. The dipolar tensor T is, in fact, determined rather more accurately for DCO than for HCO at present and can be more reliably compared with the solid state data. If Holmberg's HCO values for the principal components of T are divided by 6·514 and rotated into the inertial axis system for DCO, the predicted values of the components are $T_{aa} = 1 \cdot 95$, $T_{bb} = 0 \cdot 57$ and $T_{cc} = -2 \cdot 56$. These agree with the measured gas phase values within experimental error. However the magnitudes are really too small for an accurate comparison to be made. The Fermi contact interaction constants would normally be expected to be in the ratio of 6·514 to 1, the ratio of the magnetogyric ratios for the proton and deuteron. In fact the experimental results exhibit a small discrepancy, which might prove to be significant in a detailed discussion of the electronic structure.

There is one other interesting comparison between HCO and DCO which can be made. Curl's theory (equation (V.11)) shows that the spin-rotation tensor components for the two isotopic species should be proportional to the corresponding rotational constants. In order to make an accurate comparison we need to know precisely the relative orientations of the principal axes of the g tensor, the ε tensor for HCO and the ε tensor for DCO. The only certain piece of information is that the out-of-plane c axes are coincident. Consequently we can compare ε_{cc} values, and Curl's relationships predict that ε_{cc} for HCO should differ from the value in DCO ($-173 \cdot 9$ MHz) by the ratio of the C_0 rotational constants. The predicted value of ε_{cc} is then $-207 \cdot 7$ MHz, which agrees remarkably well with the measured value of $-207 \cdot 7$ MHz! It should be pointed out that these comparisons are based on the assumption that replacement of the proton by the deuteron does not significantly affect the electronic wave function.

There are still many measurements to be made for both HCO and DCO, but they are rewarding species to study, and the results should provide answers to a number of interesting questions of general importance.

V.7. Microwave and Electron Resonance Studies of NO_2 and ClO_2

NO_2 and ClO_2 are relatively stable molecules which possess one unpaired electron, and consequently they have been studied extensively by high-resolution spectroscopy, particularly NO_2. ClO_2 has a disconcerting habit of exploding violently, which may have put some physicists off!

The microwave studies of NO_2 fall into three main categories. Conventional microwave spectra at frequencies ranging from 15 GHz to 178 GHz have been recorded, the most recent work being by Bird, Baird, Jache, Hodgeson, Curl, Kunkle, Bransford, Rastrup-Andersen and Rosenthal (1964) and by Lees, Curl and Baker (1966). Laser electron resonance spectra have been described by Curl, Evenson and Wells (1972) at frequencies of 890·8 GHz and 964·3 GHz, and the magnetic dipole electron resonance spectrum at 9 GHz, which has been observed by many workers, has been investigated thoroughly by Schaafsma (1967).

NO_2 is a near-prolate rotor and the lower rotational levels are illustrated in Fig. V.15; this diagram is based upon the most recent values of the rotational constants, which are

$$A = 239898\cdot3 \text{ MHz}, \qquad B = 13001\cdot59, \qquad C = 12304\cdot72.$$

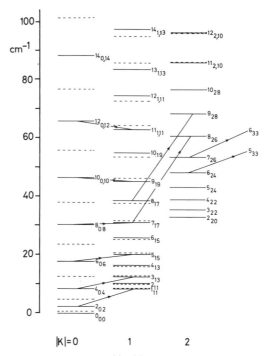

Fig. V.15. Lower rotational levels of $^{14}N^{16}O_2$ and observed transitions. The dotted lines indicate rotational levels which are absent because of nuclear spin permutation symmetry (see section III.5.5).

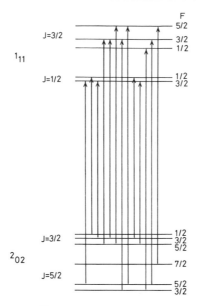

Fig. V.16. Spin-rotation and ^{14}N hyperfine splitting of the $2_{02} \rightarrow 1_{11}$ transition in $^{14}N^{16}O_2$.

The diagram does not show the spin doubling or nuclear hyperfine structure, but Fig. V.16 does illustrate these splittings in the $2_{02} \rightarrow 1_{11}$ transition, studied by Lees, Curl and Baker. The pure microwave lines which have been observed so far arise from b-type transitions between different K stacks ($\Delta K = \pm 1$) and between different N levels ($\Delta N = \pm 1$). The hyperfine structure satisfies the additional selection rules $\Delta F = 0$ or ± 1. Figure V.16 refers to the isotopic species $^{14}N^{16}O_2$; each (N, K) level is split into two electron spin components characterized by $J = N \pm \frac{1}{2}$, and the hyperfine interaction with ^{14}N results in a further triplet splitting ($F = J, J \pm 1$) except for $J = \frac{1}{2}$ which shows a doublet splitting. However several other isotopic species have also been studied, including $^{15}N^{16}O_2$ which shows a doublet hyperfine splitting from ^{15}N ($I = \frac{1}{2}$), $^{14}N^{16}O^{18}O$, and $^{14}N^{16}O^{17}O$ which shows hyperfine splitting from both ^{14}N and ^{17}O, the latter having spin $I = \frac{5}{2}$.

The laser electron resonance transitions studied by Curl, Evenson and Wells (1972) are again transitions between K stacks, namely,

$$7_{17} \rightarrow 8_{26}, \quad 8_{17} \rightarrow 9_{28}, \quad 7_{26} \rightarrow 6_{33}, \quad 6_{24} \rightarrow 5_{33}.$$

The microwave transition frequencies are close to two laser frequencies of

HCN, the small differences being compensated for by applied magnetic fields ranging from 1 to 12 kG. The applied field removes the $2F + 1$ degeneracy of each F level and the transitions studied obey the further selection rules $\Delta M_F = 0$ or ± 1. A very large number of resonance lines were detected and their measurement allows the Zeeman effect parameters to be determined, as we shall discuss later.

Finally the magnetic dipole electron resonance spectrum has been studied by Schaafsma using a conventional X-band ($\sim 9\,\text{GHz}$) electron spin resonance spectrometer. The transitions detected are, in general, $\Delta N = \Delta K = \Delta J = 0$, $\Delta M_F = \pm 1$ or in a representation in which $\mathbf{S}$ and $\mathbf{I}$ are taken to be decoupled from each other and also from $\mathbf{N}$, $\Delta M_I = 0$, $\Delta M_N = 0$, $\Delta M_S = \pm 1$. The transitions are therefore virtually pure electron spin resonance transitions,

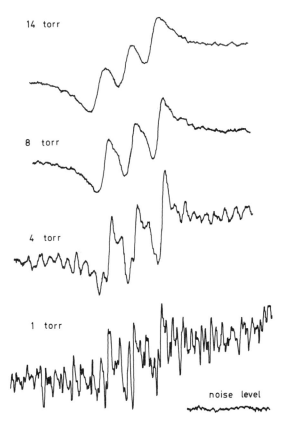

Fig. V.17. Magnetic dipole electron resonance spectrum of NO_2 at four different pressures (taken from Schaafsma (1967)).

split by nuclear hyperfine interaction, and shifted slightly by the spin-rotation interaction. Because a large number of rotational levels are populated at room temperature, the electron resonance transitions are exceedingly numerous. Schaafsma estimates that some 10^6 lines would be observable if they could be resolved; unfortunately the lines are not separated from each other and it is not possible to "analyse" the spectrum in the conventional sense. The best that can be achieved is to demonstrate that the observed pattern of lines exhibits features which are accountable in terms of the expected Hamiltonian and values of the molecular constants determined from other studies. A portion of the spectrum is shown in Fig. V.17; at relatively high pressures the spectrum exhibits three broad lines corresponding to the ^{14}N hyperfine interaction but as the pressure is lowered the splitting due to the spin-rotation coupling becomes resolved. Note that because the electron spin is effectively decoupled from the molecular framework by the magnetic field, the $\Delta M_S = \pm 1$ transitions have magnetic dipole intensity only, and a g value which is close to the free spin value. This spectrum is, in fact, characteristic of what one would observe for any non-linear doublet state species in the gas phase. For short-lived radicals, however, the lines are, in general, too weak to be observed, and even if they could be observed there would probably be too many for them to be resolved in all but the lightest species.

By combining all the microwave measurements on NO_2 it is possible to determine a considerable number of the molecular parameters. Apart from the rotational constants already quoted, the principal components of the spin-rotation tensor ε are found to be

$$\varepsilon_{aa} = 5412 \cdot 167 \text{ MHz}, \qquad \varepsilon_{aa}/A = 0 \cdot 022560$$
$$\varepsilon_{bb} = \qquad 7 \cdot 904 \qquad\qquad \varepsilon_{bb}/B = 0 \cdot 000607$$
$$\varepsilon_{cc} = -95 \cdot 524 \qquad\qquad \varepsilon_{cc}/C = -0 \cdot 007763$$

Because of the molecular symmetry the ε tensor is diagonal in the inertial axis system, which is illustrated in Fig. V.18. The Fermi contact interaction,

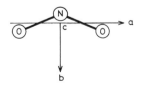

Fig. V.18. Location of the inertial axes of NO_2.

the dipolar ^{14}N hyperfine tensor, and the ^{14}N electric quadrupole tensor have also been determined, the values being,

$$a_C = 147{\cdot}23 \text{ MHz}, \qquad T_{aa} = -22{\cdot}16 \qquad (eqQ)_{aa} = 0{\cdot}45$$
$$T_{bb} = 39{\cdot}85 \qquad (eqQ)_{bb} = -1{\cdot}71$$
$$T_{cc} = -17{\cdot}69 \qquad (eqQ)_{cc} = 1{\cdot}26$$

The Zeeman effect studies of Curl, Evenson and Wells can, in principle, yield values for the components in the inertial axis system of the rotational magnetic moment tensor and the electron spin g tensor. In fact the rotational moment is rather small and therefore difficult to determine. The spin g tensor components can be determined, however, and the values given by Curl, Evenson and Wells are

$$g_{aa} = 2{\cdot}0030, \qquad g_{bb} = 2{\cdot}0205, \qquad g_{cc} = 2{\cdot}0155, \qquad g_{av} = 2{\cdot}013$$
$$\Delta g_{aa} = +0{\cdot}0007, \qquad \Delta g_{bb} = +0{\cdot}0182, \qquad \Delta g_{cc} = +0{\cdot}0132.$$

The Δg values are the differences between the measured components and the free spin value, 2·0023. Now these values can be compared with solid state measurements by Zeldes and Livingston (1961) who studied NO_2 oriented in a single crystal of $NaNO_2$. The solid state results are

$$g_{aa} = 1{\cdot}9910, \qquad g_{bb} = 2{\cdot}0015, \qquad g_{cc} = 2{\cdot}0057, \qquad g_{av} = 1{\cdot}9994,$$
$$\Delta g_{aa} = -0{\cdot}0113, \qquad \Delta g_{bb} = -0{\cdot}0008, \qquad \Delta g_{cc} = +0{\cdot}0034.$$

These do not agree with the gas phase values at all. The discrepancy between the average g values is quite large, as is also the difference in Δg values. The mystery deepens if one uses Curl's relationships (equation (V.11)) to predict the g tensor components from the gas phase measurements of the ε tensor and the rotational constants. The predicted values are

$$\Delta g_{aa} = -0{\cdot}0113, \qquad \Delta g_{bb} = -0{\cdot}0003, \qquad \Delta g_{cc} = +0{\cdot}0039.$$

These are in excellent agreement with the solid state values! It is clear that further measurements of the Zeeman effect in gaseous NO_2 are required.

The agreement between the gas phase and solid state ^{14}N hyperfine constants is fairly satisfactory, the solid state values being

$$a_C = 153{\cdot}2 \text{ MHz}, \qquad T_{aa} = -22{\cdot}3, \qquad T_{bb} = +37{\cdot}0, \qquad T_{cc} = -14{\cdot}8.$$

The ^{14}N quadrupole interaction is apparently too small to be observed and measured in the solid state spectrum.

Microwave transitions in ^{14}N^{16}O^{17}O have been studied by Foster, Hodgeson and Curl (1966); the new information arises from hyperfine interaction with ^{17}O which has $I = \frac{5}{2}$, and the hyperfine structure of the rotational transitions is extremely complex. Nevertheless it proved possible to obtain the Fermi contact interaction, the ^{17}O dipolar hyperfine coupling, and the ^{17}O electric quadrupole coupling. The constants have the following values:

$$a_C(^{17}O) = -63\cdot76 \text{ MHz,} \quad T_{aa}(^{17}O) = 50\cdot48, \quad (eqQ)_{aa} = -0\cdot202,$$
$$T_{bb} \phantom{(^{17}O)} = -106\cdot33, \quad (eqQ)_{bb} = +0\cdot211,$$
$$T_{cc} \phantom{(^{17}O)} = 55\cdot85, \quad (eqQ)_{cc} = -0\cdot009.$$

There is therefore now available a considerable amount of accurate data which contains information about the electronic wave function of NO_2. There are two approaches to the interpretation of this data. The isotropic ^{14}N and ^{17}O constants yield the s electron spin density at the respective nuclei, whilst the dipolar hyperfine constants yield values of certain averages to be computed over the electronic wave function. A computed wave function may therefore be critically examined. Alternatively, and more simply, the data can be interpreted in terms of the one-electron molecular orbital model, that is, used to provide a description of the "molecular orbital containing the unpaired electron". In terms of this model, the experimental values of the hyperfine constants show that the molecular orbital containing the unpaired electron is formed predominantly by combination of in-plane $2p$ atomic orbitals on the three atoms, the electron density on the oxygen atoms being about twice that on the nitrogen atom. The s electron densities are an order of magnitude smaller.

The g tensor components also provide information about the electronic structure of the molecule, in both the ground state and the excited states which are mixed with the ground state by spin-orbit coupling.

ClO_2 has been studied using pure microwave spectroscopy by Curl, Kinsey, Baker, Baird, Bird, Heidelberg, Sugden, Jenkins and Kenney (1961) and by Curl, Heidelberg and Kinsey (1962). Both of the predominant isotopic species, $^{35}ClO_2$ and $^{37}ClO_2$, have been investigated and the microwave spectra interpreted in terms of the following effective Hamiltonian:

$$\mathcal{H}_{\text{eff}} = AN_a^2 + BN_b^2 + CN_c^2 + \varepsilon_{aa}N_aS_a + \varepsilon_{bb}N_bS_b + \varepsilon_{cc}N_cS_c$$
$$+ a_C\mathbf{S}.\mathbf{I} + T_{aa}S_aI_a + T_{bb}S_bI_b + T_{cc}S_cI_c$$
$$+ (eqQ)_{aa}I_a^2 + (eqQ)_{bb}I_b^2 + (eqQ)_{cc}I_c^2. \tag{V.12}$$

This Hamiltonian has, of course, the same form as that used for NO_2. The lower rotational levels are shown in Fig. V.19, based upon the values of the rotational constants for $^{35}Cl^{16}O_2$,

$$A = 52077\cdot95 \text{ MHz,} \quad B = 9952\cdot42, \quad C = 8333\cdot21.$$

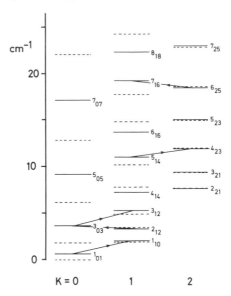

Fig. V.19. Lower rotational levels of $^{35}Cl^{16}O_2$ and observed transitions. The dotted lines indicate rotational levels which are absent because of nuclear spin permutation symmetry (see section III.5.5).

Each level shown in Fig. V.19 is split into a spin doublet, and each of these (J) levels is further split by hyperfine interaction with the chlorine, both isotopes having spin $\frac{3}{2}$ and a consequent quartet hyperfine splitting. Some of the rotational transitions which have been studied are also indicated in Fig. V.19, each being split into many components by the spin-rotation and hyperfine interactions. Measurements have been made in the frequency range 15–48 GHz and all the molecular parameters appearing in equation (V.12) have been determined for both isotopic species. For $^{35}Cl^{16}O_2$ the results are:

$$\varepsilon_{aa} = -1388 \cdot 6 \text{ MHz}, \quad \varepsilon_{aa}/A = -0 \cdot 026664, \quad T_{aa} = -77 \cdot 87, \quad (eqQ)_{aa} = -8 \cdot 65$$
$$\varepsilon_{bb} = -216 \cdot 7, \quad \varepsilon_{bb}/B = -0 \cdot 021774, \quad T_{bb} = -83 \cdot 09, \quad (eqQ)_{bb} = 0 \cdot 38$$
$$\varepsilon_{cc} = +4 \cdot 5, \quad \varepsilon_{cc}/C = +0 \cdot 000540, \quad T_{cc} = 160 \cdot 96, \quad (eqQ)_{cc} = 8 \cdot 27$$
$$a_c = 46 \cdot 21$$

It is interesting to compare these values with those for NO_2. The "reduced" spin-rotation constants are similar, except for ε_{bb}/B which is almost two orders of magnitude larger for ClO_2. The electron spin resonance spectrum of ClO_2 trapped in a crystal of $KClO_4$ has been studied by Cole (1960) and the principal components of the g tensor are found to be

$$g_{aa} = 2\cdot 0088, \qquad g_{bb} = 2\cdot 0183, \qquad g_{cc} = 2\cdot 0036$$
$$\Delta g_{aa} = 0\cdot 0065, \qquad \Delta g_{bb} = 0\cdot 0160, \qquad \Delta g_{cc} = 0\cdot 0013.$$

The predicted Δg values using Curl's relationships are

$$\Delta g_{aa} = 0\cdot 0133, \qquad \Delta g_{bb} = 0\cdot 0109, \qquad \Delta g_{cc} = 0\cdot 0003.$$

These do not agree too convincingly with the measured solid state values and again one can only conclude that gas phase measurements of the g tensor are required to clarify the position.

The dipolar hyperfine tensor components for ^{35}Cl are considerably larger than those for NO_2, and even allowance for the fact that the ^{35}Cl nuclear magnetic moment is twice as large as that of ^{14}N still reveals a substantial difference. In contrast, however, the isotropic chlorine coupling constant is much smaller than that of nitrogen. The ^{35}Cl nuclear quadrupole moment is four times larger than that of ^{14}N and opposite in sign, so that if account of this is taken, the quadrupole coupling tensors for the two molecules are similar. The solid state chlorine hyperfine constants are

$$T_{aa} = -75\cdot 3 \text{ MHz}, \qquad T_{bb} = -86\cdot 9, \qquad T_{cc} = 161\cdot 4, \qquad a_C = 43\cdot 3,$$

which agree quite well with the gas phase values. The quadrupole tensor does not, however, appear to have been determined from the solid state spectrum.

Little attempt at a quantitative interpretation of the hyperfine parameters in terms of the electronic wave function has yet been made. Using molecular orbital theory, the unpaired electron is expected to occupy a π-type molecular orbital, and the observed dipolar hyperfine constants suggest that the spin density on the chlorine atom is about 70%. The isotropic chlorine splitting indicates only a very small s electron spin density.

V.8. Concluding Remarks

Much more work on non-linear gaseous radicals is required. From the results reported so far it seems that the gas phase and solid state hyperfine constants agree reasonably well, but only for NO_2 has the electron spin g tensor been measured in both phases, with disquieting results. Moreover the expected relationship between the g tensor and spin-rotation tensor has yet to be demonstrated convincingly. It seems reasonable to suppose that electron resonance microwave spectra of other non-linear radicals, symmetric and asymmetric tops, will be detected during the next few years. Such studies will provide more accurate tests of computed electronic wave functions than are available at present.

References

Adrian, F. J., Cochran, E. L. and Bowers, V. A. (1962). *J. chem. Phys.* **36**, 1661.

Amano, T., Hirota, E. and Morino, Y. (1967). *J. phys. Soc. Japan* **22**, 399.

Amano, T., Saito, S., Hirota, E., Morino, Y., Johnson, D. R. and Powell, F. X. (1969). *J. molec. Spectrosc.* **30**, 275.

Amano, T. and Hirota, E. (1972). *J. chem. Phys.* **57**, 5608.

Anderson, J. B., Andres, R. P. and Fenn, J. B. (1966). *Adv. chem. Phys.* **10**, 275.

Arrington, C. A., Falick, A. M. and Myers, R. J. (1971). *J. chem. Phys.* **55**, 909.

Bird, G. R., Baird, J. C., Jache, A. W., Hodgeson, J. A., Curl, R. F., Kunkle, A. C., Bransford, J. W., Rastrup-Andersen, J. and Rosenthal, J. (1964). *J. chem. Phys.* **40**, 3378.

Bolman, P. H., Brown, J. M., Carrington, A. and Lycett, G. J. (1973). *Proc. R. Soc.* Ser. A. **335**, 113.

Bolman, P. H., Brown, J. M., Carrington, A. and Ramsay, D. A. (1973). *Proc. R. Soc.* Ser. A. (To be published.)

Bowater, I. C., Brown, J. M. and Carrington, A. (1971). *J. chem. Phys.* **54**, 4957.

Bowater, I. C., Brown, J. M. and Carrington, A. (1973) *Proc. R. Soc.* Ser. A. **333**, 265.

Breckenridge, W. H. and Miller, T. A. (1972). *J. chem. Phys.* **56**, 465.

Brooks, P. R., Lichten, W. and Reno, R. (1971). *Phys. Rev.* **A4**, 2217.

Brown, J. M. and Uehara, H. (1972). *Molec. Phys.* **24**, 1169.

Brown, J. M., Byfleet, C. R., Howard, B. J. and Russell, D. K. (1972). *Molec. Phys.* **23**, 457.

Byfleet, C. R., Carrington, A. and Russell, D. K. (1971). *Molec. Phys.* **20**, 271.

Cahill, P. and Healy, A. (1970). *Bull. Am. phys. Soc.* **15**, 1510.

Carrington, A. and Howard, B. J. (1970). *Molec. Phys.* **18**, 225.

Carrington, A., Levy, D. H. and Miller, T. A. (1966a). *Proc. R. Soc.* Ser. A. **293**, 108.

Carrington, A., Levy, D. H. and Miller, T. A. (1966b). *Trans. Faraday Soc.* **62**, 2994.

Carrington, A., Levy, D. H. and Miller, T. A. (1967a). *Rev. scient. Instrum.* **38**, 1183.

Carrington, A., Levy, D. H. and Miller, T. A. (1967b). *J. chem. Phys.* **47**, 3801.

Carrington, A., Levy, D. H. and Miller, T. A. (1967c). *Proc. R. Soc.* Ser. A. **298**, 340.

Carrington, A., Dyer, P. N. and Levy, D. H. (1967). *J. chem. Phys.* **47**, 1756.

Carrington, A., Fabris, A. R. and Lucas, N. J. D. (1968). *J. chem. Phys.* **49**, 5545.

Carrington, A., Fabris, A. R. and Lucas, N. J. D. (1969). *Molec. Phys.* **16**, 195.

Carrington, A., Currie, G. N., Levy, D. H. and Miller, T. A. (1969). *Molec. Phys.* **17**, 535.

Carrington, A. and Lucas, N. J. D. (1970). *Proc. R. Soc.* Ser. A. **314**, 567.

Carrington, A., Dyer, P. N. and Levy, D. H. (1970). *J. chem. Phys.* **52**, 309.

Carrington, A., Currie, G. N. and Lucas, N. J. D. (1970). *Proc. R. Soc.* Ser. A. **315**, 355.

Carrington, A., Fabris, A. R., Howard, B. J. and Lucas, N. J. D. (1971). *Molec. Phys.* **20**, 961.

Chiu, L-Y Chow (1966). *Phys. Rev.* **145**, 144.

Churg, A. and Levy, D. H. (1970). *Astrophys. J.* **162**, L161.

Clough, P. N., Curran, A. H. and Thrush, B. A. (1971). *Proc. R. Soc.* Ser. A. **323**, 451.

Cole, T. (1960). *Proc. natn. Acad. Sci. U.S.A.* **46**, 506.

Curl, R. F., Kinsey, J. L., Baker, J. G., Baird, J. C., Bird, G. R., Heidelberg, R. F., Sugden, T. M., Jenkins, D. R. and Kenney, C. N. (1961). *Phys. Rev.* **121**, 1119.

Curl, R. F., Heidelberg, R. F. and Kinsey, J. L. (1962). *Phys. Rev.* **125**, 1993.

Curl, R. F. (1965), *Molec. Phys.* **9**, 585.

Curl, R. F., Evenson, K. M. and Wells, J. S. (1972). *J. chem. Phys.* **56**, 5143.

Curran, A. H., MacDonald, R. G., Stone, A. J. and Thrush, B. A. (1973). *Proc. R. Soc.* Ser. A. **332**, 355.

Daniels, J. M. and Dorain, P. B. (1966). *J. chem. Phys.* **45**, 26.

Dehmelt, H. G. and Jefferts, K. B. (1962). *Phys. Rev.* **125**, 1318.

Dirac, P. A. M. (1928). *Proc. R. Soc.* Ser. A. **117**, 610.

Dixon, R. N. (1960). *Phil. Trans. R. Soc.* Ser. A. **252**, 165.

Dousmanis, G. C., Sanders, T. M. and Townes, C. H. (1955). *Phys. Rev.* **100**, 1735.

Evenson, K. M., Dunn, J. and Broida, H. P. (1964). *Phys. Rev.* **136**, 1566.

Evenson, K. M. (1968). *Appl. Phys. Lett.* **12**, 253.

Evenson, K. M., Broida, H. P., Wells, J. S., Mahler, R. J. and Mizushima, M. (1968). *Phys. Rev. Lett.* **21**, 1038.

Evenson, K. M. (1969). *Phys. Rev.* **178**, 1.

Evenson, K. M., Radford, H. E. and Moran, M. M. (1971). *Appl. Phys. Lett.* **18**, 426.

Field, R. W. and Bergeman, T. H. (1971). *J. chem. Phys.* **54**, 2936.

Foldy, L. L. and Wouthuysen, S. A. (1950). *Phys. Rev.* **78**, 29.

Fontana, P. R. (1962). *Phys. Rev.* **125**, 220.

Foster, P. D., Hodgeson, J. A. and Curl, R. F. (1966). *J. chem. Phys.* **45**, 3760.

Freund, R. S. and Klemperer, W. A. (1965). *J. chem. Phys.* **43**, 2422.

Freund, R. S. and Miller, T. A. (1972). *J. chem. Phys.* **56**, 2211.

Freund, R. S., Miller, T. A., De Santis, D. and Lurio, A. (1970). *J. chem. Phys.* **53**, 2290.

Freund, S. M., Herbst, E., Mariella, R. P. and Klemperer, W. A. (1972). *J. chem. Phys.* **56**, 1467.

Frosch, R. A. and Foley, H. M. (1952). *Phys. Rev.* **88**, 1337.

Gammon, R. H., Stern, R. C. and Klemperer, W. A. (1971). *J. chem. Phys.* **54**, 2151.

Gammon, R. H., Stern, R. C., Lesk, M. E., Wicke, B. G. and Klemperer, W. A. (1971). *J. chem. Phys.* **54**, 2136.

German, K. R. and Zare, R. N. (1969). *Phys. Rev. Lett.* **23**, 1207.

German, K. R., Bergeman, T. H., Weinstock, E. M. and Zare, R. N. (1973) (To be published).

Herzberg, G. (1945). "Infrared and Raman Spectra of Polyatomic Molecules" Van Nostrand Company, Princeton, N.J.

Herzberg, G. (1950). "Spectra of Diatomic Molecules" Van Nostrand Company, Princeton, N.J.

Herzberg, G. (1966). "Electronic Spectra and Electronic Structure of Polyatomic Molecules" Van Nostrand Reinhold Company, New York.

Herzberg, G. and Ramsay, D. A. (1955). *Proc. R. Soc.* Ser. A. **233**, 34.

Hinkley, R. K., Walker, T. E. H. and Richards, W. G. (1973). *Proc. R. Soc.* Ser. A. **331**, 553.

Holmberg, R. W. (1969). *J. chem. Phys.* **51**, 3255.

Huo, W. (1966). *J. chem. Phys.* **45**, 1554.

Jefferts, K. B. (1969). *Phys. Rev. Lett.* **23**, 1476.

Johnson, D. R. (1973), private communication.

Johns, J. W. C., Priddle, S. H. and Ramsay, D. A. (1963). *Discuss. Faraday Soc.* **35**, 90.

Jost, R., Marechal, M. A. and Lombardi, M. (1972). *Phys. Rev.* **A5**, 740.

Kewley, R., Sastry, K. V. L. N., Winnewisser, M. and Gordy, W. (1963). *J. chem. Phys.* **39**, 2856.

King, G. W., Hainer, R. M. and Cross, P. C. (1943). *J. chem. Phys.* **11**, 27.

Kraus, J. D. (1966). "Radio Astronomy" McGraw-Hill Book Company, New York.

Lee, K. P., Tam, W. G., Larouche, R. and Woonton, G. A. (1971). *Can. J. Phys.* **49**, 2207.

Lees, R. M., Curl, R. F. and Baker, J. G. (1966). *J. chem. Phys.* **45**, 2037.

Levy, D. H. (1972), private communication.

Lichten, W. (1960). *Phys. Rev.* **120**, 848.

Lichten, W. (1962). *Phys. Rev.* **126**, 1020.

Lide, D. R., Jr., (1963). *J. chem. Phys.* **38**, 2027.

Marechal, M. A., Jost, R. and Lombardi, M. (1972). *Phys. Rev.* **A5**, 732, 740.

McGurk, J., Tigelaar, H. L., Rock, S. L., Norris, C. L. and Flygare, W. H. (1973). *J. chem. Phys.* **58**, 1420.

Miller, T. A. (1971a). *J. chem. Phys.* **54**, 330.

Miller, T. A. (1971b). *J. chem. Phys.* **54**, 1658.

Miller, T. A. and Freund, R. S. (1972). *J. chem. Phys.* **56**, 3165.

Mockler, R. C. and Bird, G. R. (1955). *Phys. Rev.* **98**, 1837.

O'Hare, P. A. G. and Wahl, A. C. (1971). *J. chem. Phys.* **55**, 666.

Penzias, A. A., Jefferts, K. B. and Wilson, R. W. (1972). *Phys. Rev. Lett.* **28**, 772.

Poole, C. P., Jr. (1967). "Electron Spin Resonance" Wiley-Interscience, New York.

Powell, F. X. and Lide, D. R. (1964). *J. chem. Phys.* **41**, 1413.

Powell, F. X. and Lide, D. R. (1966). *J. chem. Phys.* **45**, 1067.

Pratt, D. W. and Broida, H. P. (1969). *J. chem. Phys.* **50**, 2181.

Radford, H. E. (1961). *Phys. Rev.* **122**, 114.

Radford, H. E. (1962). *Phys. Rev.* **126**, 1035.

Radford, H. E. (1964a). *J. chem. Phys.* **40**, 2732.

Radford, H. E. (1964b). *Phys. Rev.* **136**, A15.

Radford, H. E. (1968). *Rev. scient. Instrum.* **39**, 1687.

Radford, H. E. (1972), private communication.

Rao, V. M., Curl, R. F., Timms, P. L. and Margrave, J. L. (1965). *J. chem. Phys.* **43**, 2557.

Ray, B. S. (1932). *Z. Phys.* **78**, 74.

Renner, R. (1934). *Z. Phys.* **92**, 172.

Saito, S. (1970). *J. chem. Phys.* **53**, 2544.

Saito, S. (1973). (To be published.)

Saito, S. and Amano, T. (1970). *J. molec. Spectrosc.* **34**, 383.

Saito, S. and Takagi, K. (1972). *Astrophys. J.* **175**, L47.

Schaafsma, T. J. (1967). *Chem. Phys. Lett.* **1**, 16.

Schwinger, J. (1948). *Phys. Rev.* **73**, 416.

Schwinger, J. (1949). *Phys. Rev.* **76**, 790.

Silvers, S. J., Bergeman, T. H. and Klemperer, W. A. (1970). *J. chem. Phys.* **52**, 4385.

Stern, O. and Gerlach, W. (1924). *Annl. Phys.* **74**, 673.

Stern, R. C., Gammon, R. H., Lesk, M. E., Freund, R. S. and Klemperer, W. A. (1970). *J. chem. Phys.* **52**, 3467.

Takagi, K. and Saito, S. (1972). *J. molec. Spectrosc.* **44**, 81.

Ter Meulen, J. J. and Dymanus, A. (1972). *Astrophys. J.* **172**, L21.

Tinkham, M. and Strandberg, M. W. P. (1955). *Phys. Rev.* **97**, 937; **97**, 951.

Townes, C. H. and Schawlow, A. L. (1955). "Microwave Spectroscopy" McGraw-Hill Book Company, New York.

Uehara, H. (1971). *Molec. Phys.* **21**, 407.

Van Vleck, J. H. (1951). *Rev. mod. Phys.* **23**, 213.

Wahl, A. C. (1972). (Unpublished results.)

Wang, S. C. (1929). *Phys. Rev.* **34**, 243.

Wells, J. S. and Evenson, K. M. (1970). *Rev. scient. Instrum.* **41**, 226.

Wicke, B. G., Field, R. W. and Klemperer, W. A. (1972). *J. chem. Phys.* **56**, 5758.

Winnewisser, M., Sastry, K. V. L. N., Cook, R. L. and Gordy, W. (1964). *J. chem. Phys.* **41**, 1687.

Zeldes, H. and Livingston, R. (1961). *J. chem. Phys.* **35**, 563.

Author Index

Numbers in italics indicate those pages where references are given in full.

259

Subject Index

DATE DUE
